Samira Mohamady

Uncertainties of Transfer Path Analysis
and Sound Design for Electrical Drives

Logos Verlag Berlin GmbH

λογος

Aachener Beiträge zur Technischen Akustik

Editor:
Prof. Dr. rer. nat. Michael Vorländer
Institute of Technical Acoustics
RWTH Aachen University
52056 Aachen
www.akustik.rwth-aachen.de

Bibliographic information published by the Deutsche Nationalbibliothek

The Deutsche Nationalbibliothek lists this publication in the Deutsche Nationalbibliografie; detailed bibliographic data are available in the Internet at http://dnb.d-nb.de .

D 82 (Diss. RWTH Aachen University, 2016)

ISBN 978-3-8325-4431-7
ISSN 1866-3052
Vol. 26

Logos Verlag Berlin GmbH
Comeniushof, Gubener Str. 47,
D-10243 Berlin
Tel.:   +49 (0)30 / 42 85 10 90
Fax: +49 (0)30 / 42 85 10 92
http://www.logos-verlag.de

UNCERTAINTIES OF TRANSFER PATH ANALYSIS

AND SOUND DESIGN

FOR

ELECTRICAL DRIVES

Von der Fakultät für Elektrotechnik und Informationstechnik der

Rheinischen-Westfälischen Technischen Hochschule Aachen

zur Erlangung des akademischen Grades eines

Doktors der Ingenieurwissenschaften

genehmigte Dissertation

vorgelegt von

Master of Science

Samira Mohamady

aus Teheran, Iran

Berichter:

Universitätsprofessor Dr. rer. nat. Michael Vorländer

Universitätsprofessor Dr.-Ing. habil. Dr. h. c. Kay Hameyer

Diese Dissertation ist auf den Internetseiten der Hochschulbibliothek online verfügbar.

To

my mother and father

Shahin Mohammadi and Mohammad Ali Mohammady

and

my sister and brothers

Abstract

The objective of this thesis is to examine the influence of possible uncertainties during the measurement of transfer paths in systems with vibro-acoustic properties such as machines or vehicles. The transfer path could be an acoustic, structural or complex vibro-acoustic system. Uncertainties can occur due to internal and external factors during the measurement process. The trend of uncertainty propagation in sampled acoustic and structural transfer paths is studied both analytically and experimentally. This dissertation investigates an uncertainty analysis in an acoustic system with specific regard to the sensor positioning and temperature, which can be defined as external sources of uncertainties and deviations in damping factors as an internal deviation factor. In a structural system, the transfer path of the system is calculated with sensor positioning uncertainties. In each case, the relative standard deviation of the transfer paths, in comparison with the reference sample, is calculated to assess the uncertainty propagations. A statistical approach is applied to the system in the higher frequencies to evaluate the error propagation. In particular, the influence of tonal input excitation on the propagation of uncertainties is modeled to determine the behavior of the system. The main result is that, with sensor displacement of up to 6% of the geometrical length of the system from the original measured position, the deviation of the transfer path remains below $4\ dB$. The dissertation also investigates the contribution of uncertainties to sound design parameters with four sources of input excitations. The objective assessment involves the calculation of the psychoacoustic parameters, and subjective evaluation is performed with a listening test. As a final investigation, the uncertainty analysis is applied to the measurement of the transfer path in an electric vehicle. The error analysis is discussed based on the state of the maximum relative error.

Contents

List of Figures

List of Tables

1

Introduction

In vehicle design, noise and vibration are two important quality factors that are linked to the reliability and comfort of the use of the product. An essential step in noise and vibration analysis is the recognition of noise sources, as well as energy traveling paths regarding the target receiver. The transfer of sound pressure from the noise source to the target receiver is characterized by the "transfer path". This analysis has been known transfer path analysis (TPA) since the early nineties.

Transfer path measurement involves the arrangement of a measurement chain, the positioning of the sensors and actuators, and the adjustment of the excitation force, etc. In vehicle acoustics, power train and driveline mounts are the main sources of noise, and a transfer path measurement could be arranged from the engine excitation to the passenger compartment. In this case, the actuator is a shaker that implies the excitation of the engine on the structure of the system. The sensor is usually a microphone or a dummy head placed inside the vehicle cabin to measure the sound pressure level. Thus, the total interior acoustic pressure inside the vehicle can be represented by the super-positioning of different transfer paths from the source excitations to the receiver (DE VIS and HENDRICX, 1992).

TPA identifies the dynamic response of the vibro-acoustic system in the frequency domain; the dynamic response has uncertain or random properties due to a variety of sources of uncertainties. For instance, variations can be created due to imperfect measurements, environmental changes, or natural variability of the process. The propagation of error in the complex vibro-acoustic system can be predicted via sensitivity analysis, direct uncertainty assessment, Monte Carlo simulation, or Taylor series expansion, (SMITH, 1999). An example of uncertainty in transfer path measurement is shown in Figure 1.1.

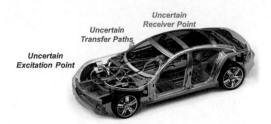

Figure 1.1: A sample of uncertainty recognition in transfer path analysis

A complete measurement involves the determination of the measured value and assessment of the range of uncertainties. The measurement quantity could be a single value or several quantities with deterministic or non-deterministic sources of uncertainties.

W. Lewin states:

"Any measurement without knowledge of uncertainties is meaningless."

The uncertainty analysis of measurement can be obtained according to the GUM, 2008 guidelines. The guidelines are very effective in defining a single value measurement quantity with known sources of measurement uncertainty. In terms of several measurement quantities and unknown sources of uncertainties, round-robin tests for measurement are applied. Using this method, several groups can measure an object and compare the results. This method provides an overall deviation of measurement quantities, without providing information on systematic error. This method is very time consuming and the results are based on statistical deviation due to unknown sources of uncertainties.

DIETRICH, 2013, addressed the uncertainty present within the primary measurement quantity, that is, the transfer function. This provides a detailed uncertainty analysis with an assessment of the independent sources of uncertainties in a measurement chain, with a focus on fundamental signal processing. His analysis aimed to predict the artifacts in the measured transfer function due to nonlinearities in the acoustic measurement chain. It also provides analytical models for airborne application by incorporating the directivity and orientation of the sound source, and the position of the source and receiver.

Insights into the trend of the uncertainty propagation along the transfer path of the system are representative of the deviation expectation. The trend analysis can be performed with a relative deviation analysis of the transfer path in terms of input uncertainty parameters. Moreover, the relative uncertainty analysis facilitates the dimensionless analysis while generalizing the expected deviation in the normalized frequency range of interest.

Besides the relative deviation of the transfer path with uncertainty parameters, the type of the excitation source impacts the uncertainty propagation. The broadband input sources excite the whole system being tested within the entire frequency range of interest; however, the system, with a tonal input excitation, samples the specific frequencies of the transfer path, which contributes to the final measured quantity. A specimen of the vibro-acoustic system with a tonal excitation is an electric vehicle with a mounted electric engine, in which the structure of the vehicle is affected by the fundamental frequency of the engine and the associated harmonics. Considering the sound propagation inside the passenger compartment of an electric vehicle, the masking effect of the combustion engine noise has gone, and hence the interior noise is very low. However, the tonal sound caused due to the electromagnetic force of the electric motor, is dominant and produces audible tonal noise (LENNSTRÖM, 2013).

The transfer of the electric engine noise is very sensitive to the features of the propagation path, thus, it is expected to monitor a large error with uncertainties in the measurement of the transfer path. Based on this assumption, a theoretical platform has been developed to calculate the influence of the uncertainties in the measurement of the system with tonal input excitations. An overview of the theory is given here:

A sample of a transfer path is given in Figure 1.2, which shows the transfer path divided by the modal and statistical region. Accordingly, the study of uncertainties of tonal excitation must consider the fundamental differences in the statistical features of two regions. In Figure 1.3, an example is shown for frequency deviation (jitter) and how this affects the magnitude of the resulting sound pressure. From this example, a statistical investigation of the jitter effect can be developed as a function of the tone frequency and its standard deviation.

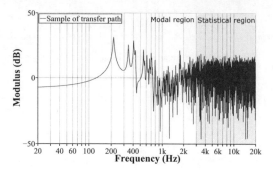

Figure 1.2: Sample of transfer path

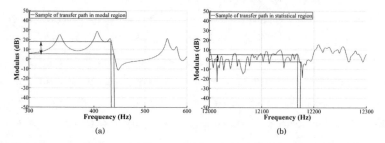

Figure 1.3: Transfer path in modal region (a), transfer path in statistical region (b) with an example of a frequency deviation (jitter) at arbitrary frequencies

In the field of vehicle acoustics, the measurement of the transfer path, with respect to input excitation, reveals the technical specifications of the running engine and structure of the vehicle under test. It also facilitates the assessment of the sound propagation pattern inside the passenger compartment for the purpose of the sound design. In the sound design process, acoustic comfort is characterized by defining the sound inside the cabin, disregarding undesired events. Sound is characterized according to the psychoacoustic parameters which are attributed to the human hearing system and psychological effects. It is expected that deviations in the transfer path of the vehicle influence the characteristics of sound field propagation in the passenger compartment.

This thesis primarily presents a relative uncertainty analysis of the transfer path, which gives insight into the trends of uncertainty propagation through the transfer path of the system being tested. For this purpose, two simple acoustic and structural systems are introduced, including input uncertainties, and the relative uncertainty analysis is performed to obtain the trend of the uncertainty propagation across the frequency range of interest. Two experimental setups are arranged to evaluate the uncertainty analysis, and the uncertainty results are compared.

Secondly, the relative uncertainty evaluation is applied to the system with tonal input excitation. Electric vehicles are the starting point for examining tonal noise; such vehicles transmit tonal engine noise through the electric power train and inside the passenger cabin. The prominent tonal noise can be intrusive and needs to be considered in the planning of the sound design of electric vehicles. On this basis, the third part of the thesis is developed to obtain the deviation of the transfer path using samples of electric engine noise and broadband noise, while also incorporating the input uncertainties. Following this, the influence of the uncertainties on the psychoacoustic parameters is calculated. As a consequence of sound design, a listening test is developed to consider the influence of uncertainties on the perception threshold of uncertainties with the included semantic test.

1.1 Objectives of this Work

In this research, the focus is

- to observe the trend of relative uncertainty propagation in the modal and statistical region of a measured transfer path in order to simplify the process of modeling uncertainties in complex structures; plus, the provision of a guideline on the maximum deviation expectation in the frequency domain.

- to reveal the influence of tonal input excitation on the behavior of the uncertainty model.

- to determine the influence of uncertainties in the measurement of the transfer path for the purpose of sound design with a variety of input excitations.

1.2 Outline of the Thesis

The following chapter summarizes the main fundamental concepts applied in this thesis. Chapter 3 is the core of the thesis and gives the analytical uncertainty analysis of the sampled transfer paths. In chapter 4, there is a discussion of the measurement setup with reference to the analytical uncertainty analysis. An evaluation of the uncertainty analysis, based on the results of the experimental measurement is put forward in chapter 5, and then the results are generalized according to the geometry size of the system to be tested. In chapter 6, the influence of the sensor positioning is assessed on the sound design parameters. This chapter presents a discussion of the objective and subjective evaluation while calculating the psychoacoustic parameters and developing a listening test. The application of relative uncertainty analysis is discussed in chapter 7, and finally, a conclusion is drawn. In the appendices, a detailed derivation is presented of some equations, as well as experimental equipment.

2

Fundamental Study

2.1 Sound and Vibration Analysis

Wave motion in the field of structure-acoustic analysis is classified according to the relation between particle motion and the direction of energy transfer. The three categories of wave motion are longitudinal, transverse, and bending waves. Longitudinal waves are generated when the particle motion and energy transfer are parallel[1], they can propagate through fluids, liquids and solids. Transverse waves are generated when the direction of particle motion and energy transfer are perpendicular, they only can be produced in solid media, Figure 2.1.

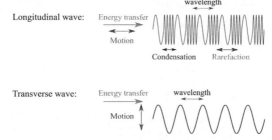

Figure 2.1: Demonstration of longitudinal and transverse wave in one-dimensional fluid and solid

The bending wave (flexural wave) propagates in the normal direction of energy transfer. It propagates through bars, beams and plates (BERANEK, 1996, FAHY et al., 2007, KUTTRUFF, 2007, KUTTRUFF, 2009, CREMER and HECKL, 2013). Propagation of longitudinal waves in fluids and bending waves in thin plates is the main focus of this research. In the next subsection, a brief review is given of the equation of wave propagation in fluids and solids.

2.1.1 Wave Propagation in Fluid

The wave equation in a rectangular coordinate system is calculated as:

$$\nabla^2 p = \frac{\rho_0}{\gamma P_0} \frac{\partial^2 p}{\partial t^2} \tag{2.1}$$

---

[1]Long in longitudinal means that the direction of the energy transfer is along the direction of wave motion.

where γ is the ratio of the specific heat of the gas at constant pressure to the specific heat at the constant volume. In this equation ρ_0 and P_0 are the density and undisturbed pressure of a fluid, respectively [2]. The wave equation in air can be expressed in a simplified form as:

$$\nabla^2 p = \frac{1}{c^2}\frac{\partial^2 p}{\partial t^2} \tag{2.2}$$

in which c is the speed of sound in air is calculated as $c = \sqrt{\frac{\gamma P_0}{\rho_0}}$. In air ($\gamma = 1.4$ with diatomic molecules), the speed of sound is about 345 $\frac{m}{s}$ at room temperature of 22°C. In problems with boundary conditions, it is useful to separate the wave equation into spatial and temporal parts (ψ and q) :

$$p = \sum_{i=1}^{n} \psi_i(r)q_i(t) \tag{2.3}$$

the spatial and temporal part of the wave are frequency dependent and the superposition of all modes(n) is required to calculate the pressure. The well-known Helmholtz equation can be derived from Equation 2.3 by introducing the wavenumber $k_i = \frac{\omega_i}{c}$ as:

$$\nabla^2 \psi_i + k_i^2 \psi_i = 0 \tag{2.4}$$

The wavenumber in the rectangular coordinates is described as:

$$k_i^2 = k_{xi}^2 + k_{yi}^2 + k_{zi}^2 \tag{2.5}$$

substituting this equation in the expanded Helmholtz equation in the Cartesian coordinate system (Equation 2.4) gives:

$$\frac{\frac{\partial X_i}{\partial x}}{X_i} + \frac{\frac{\partial Y_i}{\partial y}}{Y_i} + \frac{\frac{\partial Z_i}{\partial z}}{Z_i} + (k_{xi}^2 + k_{yi}^2 + k_{zi}^2) = 0 \tag{2.6}$$

The solution of the wave equation in each dimension:

$$\frac{\partial X_i}{\partial x} = -k_{xi}^2 X_i \implies X_i = A_i cos\left(\frac{n_{xi}\pi}{L_x}x\right) \tag{2.7a}$$

$$\frac{\partial Y_i}{\partial y} = -k_{yi}^2 Y_i \implies Y_i = B_i cos\left(\frac{n_{yi}\pi}{L_y}y\right) \tag{2.7b}$$

---

[2]The derivation of the wave equation is given in the appendix.

$$\frac{\partial Z_i}{\partial z} = -k_{zi}^2 Z_i \implies Z_i = C_i \cos\left(\frac{n_{zi}\pi}{L_z}z\right) \tag{2.7c}$$

where n_{xi}, n_{yi} and n_{zi} are the mode numbers in x, y and z direction, thus the spatial wave form is obtained as::

$$\psi_i = X_i\, Y_i\, Z_i = A_i B_i C_i \cos\left(\frac{n_{xi}\pi}{L_x}x\right)\cos\left(\frac{n_{yi}\pi}{L_y}y\right)\cos\left(\frac{n_{zi}\pi}{L_z}z\right) \tag{2.8}$$

In a bounded space, the sound field is calculated according to the specified boundaries, for instance, in a rectangular enclosure with hard boundaries, Figure 2.2, particle displacement will be zero at the boundaries and sound pressure will be at a maximum.

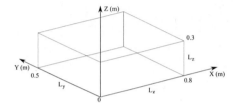

Figure 2.2: Rectangular enclosure with hard boundaries

The Equations 2.7a to 2.7c in rectangular enclosure become:

In x direction:
$$\begin{cases} \left.\frac{\partial X_i}{\partial x}\right|_{x=0} = 0; \\ \left.\frac{\partial X_i}{\partial x}\right|_{x=L_x} = 0; \end{cases} \tag{2.9a}$$

In y direction:
$$\begin{cases} \left.\frac{\partial Y_i}{\partial y}\right|_{y=0} = 0; \\ \left.\frac{\partial Y_i}{\partial y}\right|_{y=L_y} = 0; \end{cases} \tag{2.9b}$$

In z direction:
$$\begin{cases} \left.\frac{\partial Z_i}{\partial z}\right|_{z=0} = 0; \\ \left.\frac{\partial Z_i}{\partial z}\right|_{z=L_z} = 0; \end{cases} \tag{2.9c}$$

For not hard boundaries, a complex wavelength will be produced and the right-hand side values in Equation 2.9a to 2.9c should be adjusted accordingly.

2.1.2 Wave Propagation in Solids

Longitudinal, transverse and bending waves can propagate through solid mediums. The one-dimensional longitudinal wave in a bulky solid is calculated as (FAHY et al., 2007):

$$\frac{\partial^2 \xi}{\partial x^2} = \left(\frac{\rho}{B}\right)\frac{\partial^2 \xi}{\partial t^2} \Rightarrow \frac{\partial^2 \xi}{\partial x^2} = \left(\frac{1}{c_l^2}\right)\frac{\partial^2 \xi}{\partial t^2} \tag{2.10}$$

where ξ represents the displacement in x direction. Longitudinal waves are non-dispersive, since the phase speed c_l is independent of frequency ($c_l = \sqrt{\frac{B}{\rho}}$). In structures with one or two free surfaces (not constrained surfaces), longitudinal wave is limited to the quasi-longitudinal wave:

$$\frac{\partial^2 \xi}{\partial x^2} = \left(\frac{\rho}{E}\right)\frac{\partial^2 \xi}{\partial t^2} \Rightarrow \frac{\partial^2 \xi}{\partial x^2} = \left(\frac{1}{c_{ql}^2}\right)\frac{\partial^2 \xi}{\partial t^2} \tag{2.11}$$

quasi-longitudinal wave speed is $c_{ql} = \sqrt{\frac{E}{\rho}}$, and is independent of the frequency.

Transverse waves propagate through solid bodies due to shear stresses; two forms of transversal waves are plane and torsional waves. One-dimensional plane wave equation is calculated as:

$$\frac{\partial^2 \eta}{\partial x^2} = \left(\frac{\rho}{G}\right)\frac{\partial^2 \eta}{\partial t^2} \Rightarrow \frac{\partial^2 \eta}{\partial x^2} = \left(\frac{1}{c_s^2}\right)\frac{\partial^2 \eta}{\partial t^2} \tag{2.12}$$

where η is transversal displacement. The one-dimensional torsional wave equation is:

$$\frac{\partial^2 \theta}{\partial x^2} = \left(\frac{I_p}{GJ}\right)\frac{\partial^2 \theta}{\partial t^2} \Rightarrow \frac{\partial^2 \theta}{\partial x^2} = \left(\frac{1}{c_t^2}\right)\frac{\partial^2 \theta}{\partial t^2} \tag{2.13}$$

where θ, I_p and GJ are torsional displacement, polar moment of inertia per unit length and torsional stiffness of the bar, respectively.

The bending wave, which is also known as the flexural wave, is an important wave in structure acoustic interaction analysis. This wave can be matched to the amplitude of the adjacent fluid and exchange the energy to generate sound. The equation of bending motion in a bar (one-dimensional domain) can be written as:

$$EI\frac{\partial^4 \eta}{\partial x^4} = -m\frac{\partial^2 \eta}{\partial t^2} \tag{2.14}$$

here m and I represent mass per unit length of the bar and the second moment of area, respectively.

A bending wave is neither a transversal nor longitudinal wave but a hybrid of them, thus wave motion has a derivation order of four. Bending waves in infinite thin plate follows the same pattern of one-dimensional bar just as the constraint applies for the second dimension (FAHY et al., 2007):

$$\frac{EI}{1-v^2}\frac{\partial^4\eta}{\partial x^4} = -m\frac{\partial^2\eta}{\partial t^2} \tag{2.15}$$

here m is the mass per unit area of the plate and I is the second moment of area per unit width ($I = \frac{h^3}{12}$ which h is thickness of the plate). Bending stiffness of the plate is:

$$D = \frac{Eh^3}{12(1-v^2)} \tag{2.16}$$

the complete wave equation propagates through x and y direction in the plate is obtained as:

$$D\left(\frac{\partial^4\eta}{\partial x^4} + 2\frac{\partial^4\eta}{\partial x^2\partial y^2} + \frac{\partial^4\eta}{\partial y^4}\right) = -m\frac{\partial\eta^2}{\partial t^2} \tag{2.17}$$

with wavenumber of

$$k_b = \left(\frac{\omega^2 m}{D}\right)^{\frac{1}{4}} \Rightarrow c_b = \left(\omega\right)^{\frac{1}{2}}\left(\frac{D}{m}\right)^{\frac{1}{4}} \tag{2.18}$$

Bending waves are dispersive due to the dependency of the bending wave speed on the frequency.

As mentioned before, two most important wave forms in the structural-acoustic analysis are longitudinal waves in fluid and bending waves in structures. The other types of waves in solids (longitudinal, quasi-longitudinal, shear and transversal waves), apply energy internally to their medium which are beyond the scope of this thesis.

The wave speed in the air and in a thin aluminum plate are calculated and depicted in Figure 2.3. According to this plot, bending waves have lower speed in comparison with fluid at lower frequencies; however above critical frequency, f_c the bending wave is larger than the sound speed in air, the intersection point between two wave speeds is called the critical or coincident frequency. In this frequency region, the highlighted area in Figure 2.3, the plate excites the adjacent fluid, inserting extra energy to the medium.

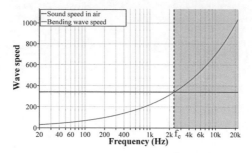

Figure 2.3: Wave speed in air and thin plate

The characteristics of the excitation source in fluid and solids are reviewed in the following subchapters.

2.1.3 Sound Source Characteristics

A longitudinal wave in three-dimensional domain propagates in a spherical shape:

$$\frac{\partial^2 p}{\partial r^2} + \frac{Z}{r}\frac{\partial p}{\partial r} = \frac{1}{c^2}\frac{\partial^2 p}{\partial t^2} \tag{2.19}$$

in comparison with the wave equation in the Cartesian coordinate (Equation 2.2), r is the radius of propagated wave, $r = (x^2 + y^2 + z^2)^{\frac{1}{2}}$. In non-reflective medium (free space), the outgoing wave is calculated as:

$$p(r,t) = \frac{A}{r}e^{j(\omega t - kr)} \tag{2.20}$$

where A is the sound pressure at unit distance from the center of the sphere, and the particle velocity in r direction is:

$$u(r,t) = \frac{A}{\rho_0 c r}\left(1 + \frac{1}{jkr}\right)e^{j(\omega t - kr)} \tag{2.21}$$

The specific impedance of wave propagation is obtained as:

$$Z = \frac{p(r,t)}{u(r,t)} = \rho_0 c\frac{jkr}{1+jkr} \Rightarrow \begin{cases} |Z| = \frac{\rho_0 c (kr)}{\sqrt{1+(kr)^2}} \\ \angle Z = \frac{\pi}{2} - tan^{-1}(kr) \end{cases} \tag{2.22}$$

For a larger distance from the spherical source $kr \gg 1$ the specific impedance can be considered as a constant, $\rho_0 c$ like plane wave, known as far field, otherwise it is a near field study should be considered. In the case of the radius of sound source, a, is smaller than one sixth of the

propagation wavelength, $ka << 1$, the sound source is considered as a point source and radiates
sound as below:

$$p(r,t) = U_0 \frac{j\omega\rho_0}{4\pi r} e^{jk(r-a)} = u_{rms} \frac{j\omega\rho_0 a^2}{r} e^{jk(r-a)} \tag{2.23}$$

where U_0 is rms volume velocity in cubic meter per second and u_{rms} is rms of source velocity.
When the sound source is not omni-directional, the spherical harmonic approach could be applied
to calculate directivity of the sound source which is much more complicated than the simple
sound source, (BERANEK, 1996, DIETRICH, 2013).

2.1.4 Structural Sources Characteristics

Structure-borne sources consist of active and passive components. The active forces are characterized by the blocked force and free velocity. The passive components are characterized
by mechanical impedance or mobility. The blocked force indicates the condition that source
impedance is theoretically infinity, which leads to zero displacement (velocity). The free velocity
can be indicated with theoretically zero impedance and consequently zero force. These two
properties of active components are connected to the passive elements as below:

$$v_f(\omega) = Y_s(\omega) f_b(\omega) \tag{2.24}$$

$$f_b(\omega) = Z_s(\omega) v_f(\omega) \tag{2.25}$$

where Z_s and Y_s are source impedance and mobility. The excitation source is described by free
velocity, v_f and blocked force, f_b.

2.2 Terminology

The following terminologies are used in this thesis:

Acoustics: is the science of generation and propagation of sound wave through elastic and
vibratory mediums (KUTTRUFF, 2007, BIERMANN, 2012). However in this thesis acoustics refers
to the science which refers to sound events in free or closed spaces.

Structural analysis: refers to the science which deals with sound events (vibration) in material
medium.

Acoustic-structural analysis: also known as vibro-acoustic analysis deals with the science of sound emission through coupled structural-acoustic systems.

Transfer path analysis (TPA): concerns propagation of sound waves through a medium from excitation source to the receiver. Throughout this thesis, the term transfer path analysis is used to refer to airborne and structure-borne transfer paths as well as coupled vibro-acoustic transfer paths. The definition of transfer paths are given in the Equation 3.10 and 3.21.

Airborne: airborne transfer path or airborne sound refers to the propagation of sound wave through air.

Structure-borne: structure-borne highlights wave propagation in the structural medium.

2.3 Expression of Uncertainty

Any measurement is subject to uncertainties. A standard method to deal with uncertainties in measurement is explained in the Guide to the Expression of Uncertainties in Measurement (GUM, 2008). According to the reference, the expected input uncertainty quantities are expressed with type A or B evaluations. Type A evaluates statistical uncertainty observations and type B discounts statistical events, like scientific judgment or previous measurement data. The standard procedure for evaluating and expressing uncertainties according to GUM summarized as (GUM, 2008):

1. Identifying the mathematical relation between measurand Y and the input quantities X_i; (model function: $Y = f(X_i)$)

2. Determining the expected value of input quantities either from measurement or external sources;

3. Evaluation of input uncertainty parameters according to Type A and Type B;

4. Obtaining measurement results and mean value y with range of uncertainty $u(y)$

5. Determining the combined standard uncertainties $u_c(y)$

6. Calculating confidence level and interval of measurement $Y = y \pm U$ for type A input quantities;

7. Calculating expanded uncertainty if necessary

The model function can be calculated or measured by using the available methods ISO, 2008. The combined uncertainty analysis is calculated based on the relation between quantities: correlated or uncorrelated.

2.3.1 Propagation of Variances

The output quantity can be represented with $y = f(x_1, x_2, ..., x_n)$, with $x_1, x_2, ..., x_n$ input quantities. Based on the assumption of the appropriate probability distribution of x_i, with the expectation of μ_i the first order Taylor series yields:

$$y - E(y) = \sum_{i=1}^{N} \frac{\partial f}{\partial x_i}\left(x_i - E(x_i)\right) \tag{2.26}$$

The assumption is that all the higher order terms are negligible. Squaring and taking the expectation of Equation 2.27 gives:

$$\sigma_y^2 = \sum_{i=1}^{N} \left[\left(\frac{\partial f}{\partial x_i}\right)^2 \sigma_{x_i}^2 + 2 \sum_{j=i+1}^{N} \frac{\partial f}{\partial x_i} \frac{\partial f}{\partial x_j} \sigma_{x_i x_j}^2 \right] \tag{2.27}$$

where σ_y^2 and σ_x^2 are variances of x and y. This equation is also called the "General law of error propagation" and for a small variation of x_i is rewritten as:

$$\Delta y = \sum_{i=1}^{N} \frac{\partial f}{\partial x_i} \Delta x_i \tag{2.28}$$

where Δy is the deviation of the output due to the small changes of input Δx_i. For a single deviation parameter it can be written as:

$$\Delta y = \frac{\partial f}{\partial x} \Delta x \tag{2.29}$$

In the decibel scale the level of deviation in amplitude is expressed as below:

$$L_{\sigma_y} = 20 log(1 + \frac{\sigma_y}{\bar{y}}) \tag{2.30}$$

where σ_y is the standard deviation of output quantity, which is referred as the mean value of the output quantity $\bar{y}$.

2.3.2 Uncertainty Modeling and Analysis

In the uncertainty model, a transfer path of the system being tested is considered as a core element and any variations in the system are introduced as input uncertainty parameters. The uncertainty model is evaluated according to the type of input excitation: broadband or tonal sources. The uncertainty analysis involves a deviation analysis with respect to the reference samples in the frequency domain as well as in the one-third octave band. The outcome of the uncertainty model is a group of responses with regard to each input uncertainty deviation. The proposed concept is shown as a block diagram in Figure 2.4

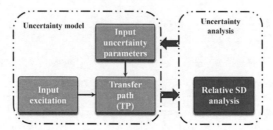

Figure 2.4: Block diagram of uncertainty analysis

Two simple acoustic and structural systems with complex transfer paths are introduced to perform the uncertainty analysis. These systems create the complex transfer path due to the large deviation with the high slope in the modal region. It should be mentioned that the modeling in this thesis follows the assumption of linearity and reciprocity. The modeling of each system is explained specifically in the next chapter.

3

Analytical Uncertainty Analysis of Transfer Path

The main focus of this chapter is uncertainty modeling and analysis of the transfer path. In this respect, firstly, the concept of transfer path analysis (TPA) is briefly explained and the uncertainty modeling approach is introduced. The modeling procedure for two simple acoustic and structural transfer paths is included and the sources of possible uncertainty quantities are determined. Consequently, a Monte Carlo simulation was used to simulate the uncertainty model. In addition to the broadband frequency analysis of the uncertainty model, the influence of tonal input excitation on the uncertainty model is discussed. Finally, a statistical procedure is applied to determine the significant differences of the transfer path with uncertainties in the higher frequencies.

3.1 Concept of Transfer Path Analysis

3.1.1 Transfer Path Analysis (TPA)

Transfer path analysis (TPA) is a well-known method of tracing the flow of air and structure-borne energy from the excitation sources to the defined receiver in the frequency domain. In general, vibro-acoustic systems consist of two parts, namely, active and passive components. Volume velocity and force excitation (Q_i, F_j) are active parts due to adding extra energy to the system and passive parts are receiver points; sound pressure (p_k) and acceleration (a_l), and interfaces between excitation sources to receivers (TP_{ik}, TP_{jk}). The proposed concept is shown as a block diagram in Figure 3.1:

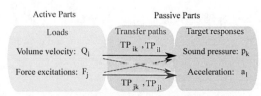

Figure 3.1: Block diagram of Transfer Path Analysis

The block diagram can be concluded with the following equations:

$$p_k(\omega) = \sum_{i=1}^{n} TP_{ik}(\omega)Q_i(\omega) + \sum_{j=1}^{m} TP_{jk}(\omega)F_j(\omega) \tag{3.1}$$

$$a_l(\omega) = \sum_{i=1}^{n} TP_{il}(\omega)Q_i(\omega) + \sum_{j=1}^{m} TP_{jl}(\omega)F_j(\omega) \tag{3.2}$$

in which TP_{ik}, TP_{il} indicate the acoustic and structural transfer paths through volume velocity excitation, and TP_{jl}, TP_{jk} indicates the acoustic and structural transfer paths through the force excitation (GAJDÁTSY, 2011).

3.1.2 Input Excitation

The input source could have broadband or tonal characteristics. In the case of broadband sources, the passive part of the system gets energy in the entire frequency range, while with tonal source the specific parts of the system contribute to the final response. Since this study is concerned with the influence of the tonal sources, a sample of tonal noise is calculated below:

$$S_t(f) = \sum_{n=1}^{N} A_n \delta(f - nf_0) + Noise \tag{3.3}$$

where f_1 is the first harmonic of the tonal excitation and f_n is the nth order of the tonal excitation. The tonal excitation with white noise up to the 10th order is illustrated in Figure 3.2.

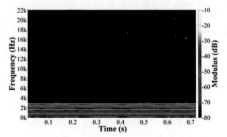

Figure 3.2: Spectrogram of tonal excitation with added white noise

In electric vehicles, the electric engine creates tonal noise. The electric engine noise is analytically calculated during the run-up in the time domain and is given in the equation 3.4, (van der GIET, 2011).

$$S_e(t) = L_p e^{-j2\pi\mu\left(n_1 + \frac{n_2-n_1}{T_{run}}t\right)t} + Noise \tag{3.4}$$

where n_1, n_2, μ and T_{run} indicate the start speed, end speed, frequency order and the run-up time, respectively.

The aerodynamic acoustic power is calculated (HECKL and MÜLLER, 1994):

$$P = \gamma_0 \left(\frac{2\pi R_r n}{c} \right)^{5.5} S \tag{3.5}$$

The mechanical speed, rotor radius and the radiating surface are indicated with n, R_r and S in equation 3.5. In this equation, γ_0 is a constant to be in a range of 500...5000 W/m^2. Finally, the sound pressure level of the simulating surface is calculated with:

$$\bar{L_P} = \left[10log\frac{\gamma_0}{I_0} + 55log\left(\frac{2\pi R_r n}{c}\right) + log\frac{S}{S_1} \right] \tag{3.6}$$

$\gamma_0 = 3.10^3$ W/m^2 gives $10log\frac{\gamma_0}{I_0} = 155dB$. S_1 is the sound pressure and is measured above the surface. A sample of engine noise during the run-up with synthetic amplitude up to the 10th order is simulated and plotted in Figure 3.3

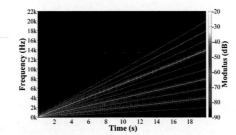

Figure 3.3: Spectrogram of simulated engine noise during run-up

The speed changes from 10 rpm to 10^3 rpm with maximum frequency of 22 kHz. The engine noise is calculated up to the 10th harmonic. In the simulation, the aliasing effect is considered as well.

3.2 Transfer Path Modeling and Analysis

3.2.1 Acoustic Transfer Path Modeling

The proposed acoustic model is a rectangular enclosure with the dimensions $0.8 \times 0.5 \times 0.3$ m^3. A golden ratio between the main axes of the enclosure is defined to guarantee a more regular mode distribution (DIETRICH et al., 2010). Absorption coefficients of the boundaries are set to zero to provide the hard boundary condition. The system is excited with an interior point source and the pressure is observed in the receiver point (Figure 3.4):

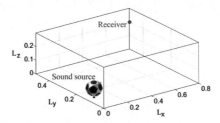

Figure 3.4: Rectangular enclosure with sound source and receiver

The acoustic pattern inside the enclosure is calculated via a modal synthesis concept (*Springer Handbook of Acoustics*; KUTTRUFF, 2009). Referring back to the Equation 2.8, the wave synthesis (eigenfunction) superposition at the sound source and receiver position is calculated with:

$$\psi_i(x,y,z) = A_i B_i C_i cos\left(\frac{n_{xi}\pi}{L_x}x\right)cos\left(\frac{n_{yi}\pi}{L_y}y\right)cos\left(\frac{n_{zi}\pi}{L_z}z\right) \tag{3.7}$$

where n_{xi}, n_{yi} and n_{zi} represent the modal numbers in x, y and z directions, respectively. In the following equations, the eigenfunction at the source is defined with $\psi_i(r_0)$ and at the receiver point with $\psi_i(r)$. To calculate the spatial pattern of the acoustic transfer path inside the enclosure, the integral is required over product of two wave patterns. According to the orthogonality condition, the integral over the product of the two dissimilar modes in the enclosure volume is zero and in two similar modes is equal to unity, thus:

$$\int_V \psi_m(r)\psi_n(r)dV = \begin{cases} 0 & m \neq n \\ V & m = n \end{cases} \tag{3.8}$$

in this equation, V indicates the volume of the enclosure. The eigenfrequencies of the enclosure (standing wave) are calculated with:

$$\omega_i = c\pi\sqrt{\left(\frac{n_{xi}}{L_x}\right)^2 + \left(\frac{n_{yi}}{L_y}\right)^2 + \left(\frac{n_{zi}}{L_z}\right)^2} \tag{3.9}$$

The sound speed is indicated with c in this equation. Eventually, the acoustic transfer path between the sound source and receiver inside the enclosure is derived from KUTTRUFF, 2009:

$$TF(\omega) = \frac{p(\omega)}{Q} = -\frac{4\pi c^2}{V} \sum_i \frac{\psi_i(r)\psi_i(r_0)}{(\omega^2 - \omega_i^2 - 2j\delta_i\omega_i)K_i} \tag{3.10}$$

where K_i is a normalized constant to include the orthogonality condition (Equation 3.8). The modal damping constant is denoted with δ_i and is calculated by means of modal reverberation time RT_i inside the scenario:

$$\delta_i = \frac{3ln(10)}{RT_i} \tag{3.11}$$

where i represents the mode number. The number of modes in the enclosure is calculated with (KUTTRUFF, 2009):

$$N(f_{max}) = \frac{4\pi}{3}V\left(\frac{f_{max}}{c}\right)^3 + \frac{\pi}{4}S\left(\frac{f_{max}}{c}\right)^2 + \frac{L}{8}\left(\frac{f_{max}}{c}\right) \approx \frac{4\pi}{3}V\left(\frac{f_{max}}{c}\right)^3 \tag{3.12}$$

The modal density is calculated by the derivative of the mode number with respect to the frequency:

$$\frac{dN}{df} = \frac{4\pi}{c^3}V\left(f_{max}\right)^2 + \frac{\pi}{2c^2}S\left(\frac{f_{max}}{c}\right) + \frac{L}{8} \approx \frac{4\pi}{c^3}V\left(f_{max}\right)^2 \tag{3.13}$$

Three types of modes can be extracted from the modal analysis of the three-dimensional system:

1. **Axial modes:** there are no modes in two axes of the system

2. **Tangential modes:** there are no modes in one axis of the system

3. **Oblique modes:** all the axes involve modes.

The axial, tangential and obliques modes as well as total number of modes of the enclosure are plotted in Figure 3.5.

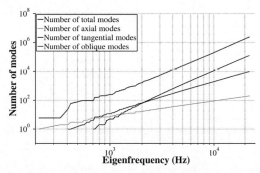

Figure 3.5: Number of total, axial, tangential and oblique modes in the enclosure

The higher number of modes fits with the length of the geometry in the higher frequencies. The overlap of modes happens when on average three eigenfrequencies fall into one resonance half-width. This frequency is called the Schroeder frequency which is named after SCHROEDER, 1962 and is calculated as below:

$$\langle \Delta f_i \rangle = 3 \left(\frac{1}{dN/df} \right) \tag{3.14}$$

The half-width of the resonance curve is calculated (SCHROEDER, 1962, KUTTRUFF, 2009):

$$\langle \Delta f_i \rangle = \frac{\langle \delta_i \rangle}{\pi} \tag{3.15}$$

where f_i indicates the ith eigenfrequency. Thus according to the definition of the Schroeder frequency, it is derived here:

$$f_{Schroeder} = \sqrt{\frac{3c^3}{4V\delta}} \tag{3.16}$$

The Schroeder frequency can be calculated by substituting the damp constant from Equation 3.11 into Equation 3.16:

$$f_{Schroeder} = \sqrt{\frac{3c^3}{4V} \left(\frac{RT}{1n(10)} \right)} = 2000 \sqrt{\frac{RT}{V}} \tag{3.17}$$

the Schroeder frequency is about 3.6 kHz for the suggested acoustic scenario with the average reverberation time of 0.4 s. A transfer function between sound source and receiver based on equation 3.10 is plotted in Figure 3.6. The first eigenfrequency ($f_1 = 232Hz$) of the enclosure and the Schroeder frequency ($f_s = 3.6kHz$) are given in Figure 3.6, respectively.

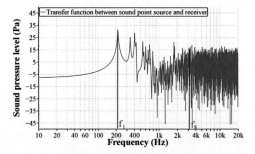

Figure 3.6: Sample transfer path

3.2.2 Structural Transfer Path Modeling

The structural transfer path is modeled using a simplified structure, the structure consists of a rectangular thin aluminum plate with simply supported edges (SSSS). The plate dimensions are $0.8 \times 0.5 \times 0.001 \ m^3$ and the mass is $2.7 \ kg/m^2$. A force excitation point and a receiver are located at two diagonal opposite corners of the plate, Figure 3.7.

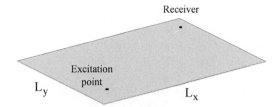

Figure 3.7: Thin aluminum plate

A bending wave propagates through the structure as a most dominant wave (Chapter 2). The eigenfunction of the plate at the excitation point and receivers is calculated with:

$$\phi_i(x,y) = 2\sin\left(\frac{n_{xi}\pi}{L_x}x\right)\sin\left(\frac{n_{yi}\pi}{L_y}y\right) \tag{3.18}$$

and, consequently, eigenfrequencies are given:

$$\omega_{pi} = \sqrt{\frac{D}{m}}\left[\left(\frac{n_{xi}\pi}{L_x}\right)^2 + \left(\frac{n_{yi}\pi}{L_y}\right)^2\right] \tag{3.19}$$

where m is the mass of plate per unit area, and D is the flexure stiffness of plate:

$$D = \frac{Eh_p^3}{12(1-v^2)} \tag{3.20}$$

with h_p the thickness of the plate. The transfer path between the excitation point and receiver is calculated (FAHY et al., 2007, CREMER and HECKL, 2013):

$$TF_p(\omega_p) = \frac{a(\omega_p)}{F} = \frac{-4\omega_p^2}{mL_xL_y}\sum_i \frac{\phi_i(x,y)\phi_i(x_0,y_0)}{\omega_{pi}^2(1+j\eta)-\omega_p^2} \tag{3.21}$$

where F is the uniform input excitation force, and η indicates damp factor, (CREMER and

HECKL, 2013):

$$\eta = \frac{f}{b} \tag{3.22}$$

whose b indicates the half-value bandwidth of the resonance curve. The orthogonality condition is also an important aspect of the following analysis:

$$\int_S m\phi_n(x,y)\phi_m(x,y) = 0; \; m \neq n \tag{3.23}$$

with S as the surface of the plate. The number of bending modes can be calculated in the frequency range of interest:

$$N_p(f_{max}) = \frac{S}{2}\sqrt{\frac{m'}{D}}f \tag{3.24}$$

the equation shows a proportional relation with the frequency, however in the acoustic system the number of modes increases with the squared of the frequency. In Figure 3.8 the total number of the modes per frequency is shown.

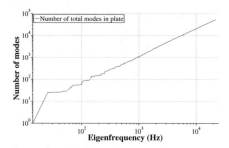

Figure 3.8: Number of total modes in the plate

Here, the response of the analysis can also be divided into two regions: the modal region, where the individual modes dominate the response of structural system, and the statistical region, where no individual modes dominate. In the literature, there is no report on the calculation of the frequency which separates the overlap domains. Thus this frequency is calculated in this research. For calculation of the overlap frequency in the thin structure, the same concept of the Schroeder frequency is considered, thus it is estimated that the maximum of three eigenfrequencies falls in one-resonance half-width:

$$\langle \Delta f \rangle = 3\left(\frac{1}{dN_p/df}\right) \tag{3.25}$$

modal density of the plate is calculated as:

$$\frac{dN_p}{df} = \frac{S}{2}\sqrt{\frac{m'}{D}}$$

(3.26)

and consequently:

$$\langle \Delta f \rangle = \frac{6}{S}\sqrt{\frac{D}{m'}}$$

(3.27)

and by considering the width of the eigenfrequency, we derive:

$$\langle \Delta f \rangle = b = \eta f$$

(3.28)

thus Equation 3.27 and 3.28 lead to:

$$\eta f = \frac{6}{S}\sqrt{\frac{D}{m'}} \Rightarrow f_S = \frac{6}{S\eta}\sqrt{\frac{D}{m'}}$$

(3.29)

the overlap frequency of the plate is calculated according to the specification of the plate as 976 Hz. The transfer function between the excitation force and receiver is demonstrated in Figure 3.9. In this graph, the overlap frequency and first eigenfrequency of the plate are highlighted.

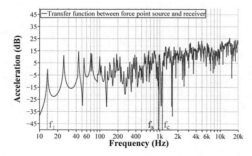

Figure 3.9: Sample transfer path

The second important aspect of the structural analysis is the critical frequency, where the air wave speed matches the wavelength of the plate, λ_b and radiates sound in its surroundings ($\lambda_b = \lambda_0 = \frac{c_0}{f_c}$). The critical frequency in the plate is calculated as (CREMER and HECKL, 2013):

$$f_c = \frac{c_0^2}{2\pi}\sqrt{\frac{m'}{D}}$$

(3.30)

The critical frequency of the specified plate is $1.2\,kHz$, Figure 3.9. Now, following the introduction of models for acoustic and vibro-acoustic transfer paths, the focus of study is on how uncertainties propagate from input parameter variation into the resulting sound pressure.

3.3 Input Uncertainty Parameters

In general, input uncertainty parameters could be divided into interior and exterior parameters. Variation in the geometry size, material, rigidity of joints are parts of the interior uncertainty parameters, while any exterior events like temperature changes or displacement in the sender or receiver points are considered as exterior uncertainty sources, (Figure 3.10).

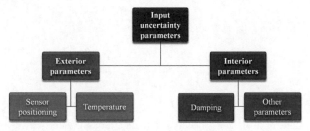

Figure 3.10: Input uncertainty parameters

It should be noted that both groups of uncertainties could be expanded in more detail. In this thesis, deviations are examined during the measurement, thus the sensor positioning and temperature variation are selected as the most dominant exterior parameters, and the decay time from the interior source of uncertainties.

In the acoustic system, all three input uncertainty parameters are calculated and discussed. In the structural analysis, uncertainties in the sensor positioning are investigated and the other two parameters are excluded due to due to redundancy. The relative standard deviation (SD) of the uncertainty model is calculated for each case according to GUM, 2008 and subsection 2.3.1.

Referring back to the block diagram of Figure 2.4, the uncertainty model is simulated by considering the influence of the input excitation and the input uncertainty parameters. The error analysis in the case of broadband excitation applies to the entire frequency range of interest, although in the case of tonal excitation some parts of the transfer path controls the deviations of the system.

The method of uncertainty analysis is explained with a simple example, which is applied to all of the uncertainty cases. It is assumed that the receiver has a small displacement from the primary position (Receiver 1), the second transfer path is calculated by adopting the equation 3.10 to give the new sensor position (Receiver 2), Figure 3.11.

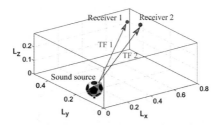

Figure 3.11: Transfer function between sound source and receiver

The calculated transfer paths are indicated in Figure 3.12

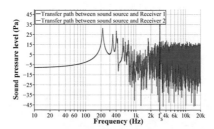

Figure 3.12: Sample of two transfer paths

The deviation of two transfer paths can be extracted by calculating a relative standard deviation (SD) between two samples:

$$L_\sigma(f) = 20\log(1 + \frac{\sigma_p(f)}{\bar{p}(f)}) \tag{3.31}$$

whose σ_p and $\bar{p}$ are SD and the mean value of the sound pressure level at each frequency bin. The result of the relative SD deviation of two sampled transfer paths is illustrated in Figure 3.13.

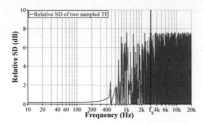

Figure 3.13: Relative SD between two sampled transfer paths

In order to obtain more information about the uncertainty analysis, the averaged power of the relative SD deviation in the one-third octave band is calculated using:

$$L_{\sigma_{1/3\,oct}} = 20\log\left[(1 + \sqrt{\sum_{1/3\,oct}\left(\frac{\sigma_p(f)}{\bar{p}(f)}\right)^2}\right] \tag{3.32}$$

The result of the averaged power of the relative SD is shown in Figure 3.14. The apparent increase of the error as the frequency increases can be captured in this analysis, while above the Schroeder frequency, $3.6\,KHz$ the error stays around 4 dB.

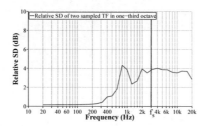

Figure 3.14: Added power of the relative SD between two sampled transfer paths in the one-third octave band

3.3.1 Acoustic Uncertainty Analysis

Sensor Positioning Uncertainties – Statistical Analysis

An earlier step before evaluating the sensor positioning uncertainty in a system, is defining the excitation source and the reference receiver position. In the measurement of the acoustic transfer paths, the exemplary system can be considered as a black box including the excitation source and the black box can be subdivided into the smaller areas (RISSLER, 2011). In Figure 3.15 the concept is shown for the enclosure example.

Figure 3.15: Black box concept

The size of the subareas can be defined according to the requested accuracy of the measurement. It is suggested to define the size of the subareas as small as the half of the examined wavelength (VOGT, 2006). The center of the subareas is a reference measurement position and the next position is at a distance of half of the wavelength.

Considering the example of the enclosure with the Schroeder frequency about 3.6 kHz, half of the wavelength is about 5 cm. Thus, based on this assumption the uncertainty analysis can be continued.

The same concept of two sampled displacements is expanded for 400 normal sampled displacements around the default receiver position. The deviation radius is about 3 cm which is less than 5 cm. Monte Carlo simulation is used to model 400 transfer functions.

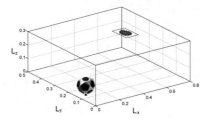

Figure 3.16: Enclosure area 1

The transfer function between the sound source and 400 receivers is plotted in Figure 3.17. The bold line represents the transfer function of sample number 200 and a cloud of light color represents the transfer functions of 400 receivers.

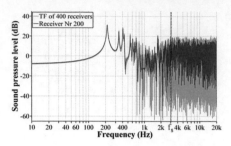

Figure 3.17: Transfer path of the enclosure with 400 receivers

The relative SD of the random positioning of 400 receivers is obtained via Equation 3.31 and the averaged power of error in the one-third octave band is computed by Equation 3.32. The results of both analyses are shown in Figure 3.18

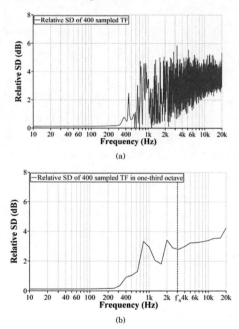

Figure 3.18: Relative SD analysis of sampled 400 TF (a), Added power of the relative uncertainties analysis in one-third octave bands (b)

Relative SD analysis indicates an increase in the frequency up to the Schroeder frequency, above this range, the deviation stays below 4 dB. It can be interpreted as saturation of the error at the higher frequencies (diffuse field). In order to examine the dependency of the analysis on the default sensor position (sample 200), three extra center receiver positions are selected inside the enclosure and 400 samples are taken from each default position. The uncertainty analysis method is applied to the three new sampled positions inside the enclosure. All four areas are shown in Figure 3.19

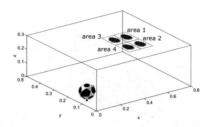

Figure 3.19: Enclosure with four measurement areas

The relative SD deviation of each case is calculated and plotted in Figure 3.20:

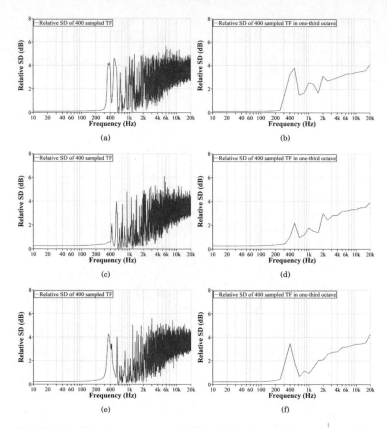

Figure 3.20: Relative SD analysis of sampled 400 TF in: area 2 (a), area 3 (c), area 4 (e) and the averaged power of the relative uncertainties analysis in one-third octave bands in: area 2 (b), area 3 (d), area 4 (f)

In all cases, the increase of error can be captured, however, the shape of the deviation in the middle frequency is different due to the specific modal responses at the receiver region, and yet the tendency of the error analysis is uniform. This conforms to the saturation at the higher frequencies (Figure 3.20(b), (d) and (f)). Now, the question is how the error propagates through the frequency domain with the increasing size of sensor displacement. In this respect, four spatial propagation spheres are defined inside the box with the growing radius. The propagation of receivers in the imaginary spheres is indicated in Figure 3.21.

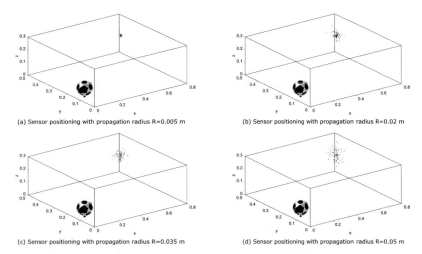

(a) Sensor positioning with propagation radius R=0.005 m

(b) Sensor positioning with propagation radius R=0.02 m

(c) Sensor positioning with propagation radius R=0.035 m

(d) Sensor positioning with propagation radius R=0.05 m

Figure 3.21: Enclosure with growing measurement area in 4 steps: (a) R=0.005 m, (b) R=0.02 m, (c) R=0.035 m, (d)=0.05 m

The uncertainty analysis is applied to the receiver in each case and indicated in Figure 3.22. It can be observed that the growth of uncertainties with the sensor displacement up to 6% of the geometrical length of the enclosure results in the same saturation of error at higher frequencies.

Schroeder predicted that the average range of the statistical fluctuations between maxima and minima equal to 10 db, which gives 5dB fluctuation from the mean value, thus the Schroeder theory supports the relative deviation analysis obtained in this thesis. SCHROEDER, 1996.

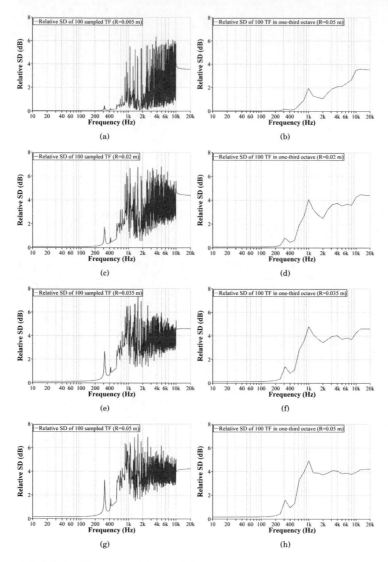

Figure 3.22: Relative SD analysis of the sampled 100 TF in: sphere 1 (a), sphere 2 (c), sphere 3 (e),
sphere 4 (g) and the averaged power of the relative uncertainties analysis in one-third
octave band in: sphere 1 (b), sphere 2 (d), sphere 3 (f), sphere 4(h)

For a complete transfer path analysis (TPA) of a given system, synthesizing of the possible

transfer paths is required from the derived sources to the receiver position.

Sensor Positioning Uncertainties with Tonal Excitation

The uncertainty model with tonal input excitation is obtained (Equation 3.3):

$$Y(f) = S_t(f) \times TF \tag{3.33}$$

The error analysis of the system excited with tonal excitation is provided below:

$$\sigma^2 = \sigma_1^2 + \sigma_2^2 + \dots + \sigma_n^2 \tag{3.34}$$

with σ_1 to σ_n are the SD of the first and n^{th} harmonic of the tonal excitation. The total error related to the input excitation is calculated with:

$$L_{\sigma_{total}} = 10log\left[1 + \sqrt{\sum_n \left(\frac{\sigma_n}{S_t(f)TF}\right)^2}\right] \tag{3.35}$$

The result of the following analysis gives a value returning the total error due to the tonal excitation. This analysis could be applied to the system with run-up engine noise excitation, Figure 3.3. During the run-up time, the harmonic frequencies of the engine noise sweeping the frequency. The maximum sweeping of the harmonics of the engine noise is limited with the run-up time and the maximum frequency. The maximum time and frequency of the electric engine in this example are defined as 5 seconds and 22 kHz, respectively. Considering the maximum frequency range of interest, the first harmonic could be increased up to 2.8 kHz, and the associated harmonics up to 22 kHz, respectively.

The uncertainty analysis of the tonal excitation in Equation 3.35 is applied to the simulated run-up engine noise and the result is plotted in Figure 3.23. The first harmonic frequency of the engine noise is defined at 100 Hz and is increased up to 2.8 kHz. The harmonics are simulated up to 10th order. The result of the following analysis is illustrated in Figure 3.23. In the system with total excitation the error increases up to 1 kHz, then the error is slightly fluctuating under 4 dB. In this figure the vertical axis is according to the shift of the first harmonic of engine noise excitation. This axis can be contributed as a time increase.

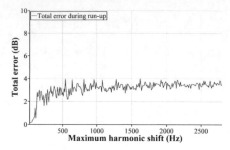

Figure 3.23: Total error due to increase of the first harmonic of tonal excitation in area 1 of the enclosure

Temperature Uncertainties in Acoustic Systems

Temperature variation occurs during the transfer path measurement which causes the changes in the speed of wave propagation in the medium and is considered as one of the sources of uncertainty. To consider the influence of temperature changes in the measured TF, the temperature varies with $\Delta T = 15°C$, and between 15 to 30 $°C$ is assumed. The transfer path due to temperature changes is calculated and presented in Figure 3.24.

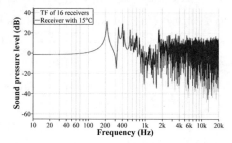

Figure 3.24: Transfer path of the enclosure with temperature uncertainties (15-30°C)

The uncertainty analysis of the transfer path according to Equation 3.31 and 3.32 is given:

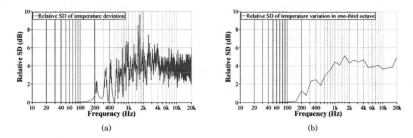

Figure 3.25: Relative SD analysis of TF with temperature changed 15 to $30°C$ (a), averaged power of the relative SD analysis in one-third octave band (b)

Above the Schroeder frequency, $3.6\ kHz$, error analysis indicates the saturation repeatedly, likewise the saturation is around $4\ dB$. It is interesting to observe that the uncertainties with $\Delta T = 15°C$ and sensor positioning leads to the same saturation deviation in this case study.

Temperature Uncertainties with Tonal Excitation

The influence of tonal excitation in the propagation of uncertainties in the system with temperature variation is calculated and plotted in Figure 3.26:

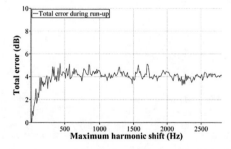

Figure 3.26: Total error due to increase of the first harmonic of tonal excitation for temperature variation

The system with temperature uncertainties and tonal excitation shows the total error around $4\ dB$ which is similar to the diffuse field.

Damping Uncertainties

In order to simulate the uncertainties in the damping of the system, an experimental setup has been arranged to measure the damp factor of the system with 17 various observation materials inside the enclosure. The experimental setup is explained in the next chapter. Since the acoustical system is the main focus of the case study, the damping of the system is analyzed via calculation of the reverberation time: Figure 3.27 shows the measured 17 cases of reverberation time in the frequency domain.

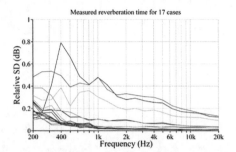

Figure 3.27: Reverberation time for 17 cases

The reverberation time in the lower frequency is model dependent, however, it is relatively constant in the diffuse field. The result of the deviation analysis according to Equation 3.31 and 3.32 is given:

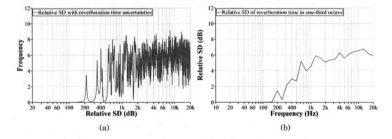

Figure 3.28: Relative SD analysis of TF with 17 cases of reverberation time (a), averaged power of the relative SD in one-third octave band (b)

The analysis also shows an increase in the frequency domain, however, the saturations occur at higher amplitude, $6\ dB$. It can be observed that the saturation starts from $1\ kHz$ which is lower than the Schroeder frequency. This can be attributed to the measured reverberation time which shows a stable behavior above $1\ kHz$. The materials inside the enclosure lead to the higher observations that create the early diffuse field. Thus it can be concluded that, as an external source of uncertainties, damping has the greatest influence in comparison with the external uncertainty source like sensor positioning and temperature.

Damping Uncertainties with Tonal Input Excitation

The influence of the tonal excitation on the system is analyzed by varying the damping factor..
The total error is calculated according to Equation 3.35 and the result is presented in the next
figure:

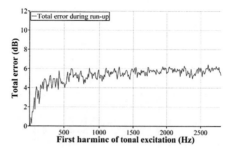

Figure 3.29: Total error due to increase of the first harmonic of tonal excitation for reverberation time
variations

The analysis shows a uniform error excitation above $1\ kHz$ and below $1\ kHz$ it is increasing
as indicated by the gently rising slope (see Figure 3.29). The total error with the tonal excitation
shows the error of $10\ dB$.

Combined Reverberation Time and Sensor Positioning Uncertainties

The external and internal sources of uncertainties in this example are uncorrelated. According
to the GUM, 2008 the uncorrelated input quantities could be calculated as below:

$$u_c^2(y) = \sum_{i=1}^{N} u_i^2(y) \tag{3.36}$$

where $u_i(y)$ represents the uncertainty parameters and $u_c(y)$ the combined uncertainties. In
the case of calculated averaged power of error in the one-third octave band (Equation 3.32) the
combined uncertainty analysis is calculated as below:

$$L_{\sigma_c} = 10\log\left[1 + \sum_{i=1}^{N}\left(\sum_{1/3\ oct}\left(\frac{\sigma_{p_i}(f)}{\bar{p}_i(f)}\right)^2\right)\right] \tag{3.37}$$

3.3.2 Structural Uncertainty Analysis

The structural system is the subject of interior and exterior uncertainties. Here, as an example
of external uncertainties, the sensor positioning uncertainties are introduced to the structural

system. This is presented in the next sub-section.

Sensor Positioning Uncertainties in the Structural System

The Monte Carlo simulation has been used to describe the analytical model for the uncertainties in sensor positioning in a thin plate Figure 3.30.

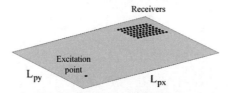

Figure 3.30: Uncertainty modeling of plate

The transfer path of the structural system is plotted in Figure 3.31:

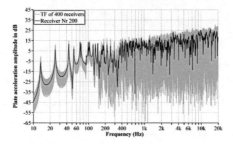

Figure 3.31: Uncertainty modeling of plate

The relative SD analysis is performed according to Equation 3.31 and 3.32, and plotted in Figure 3.32.

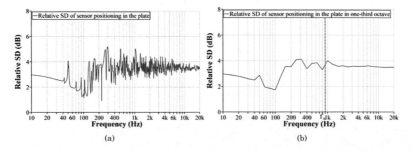

Figure 3.32: Relative SD analysis of sampled 400 TF of the plate (a), added power of the relative uncertainties analysis in one-third octave bands in the plate (b)

The relative SD analysis shows a relatively constant deviation of the structure in respect of the sensor positioning uncertainties, while above the overlap frequency of the uncertainty is constant. This means that the structural system is very sensitive to the sensor positioning in general and produces higher deviation in the entire frequency range.

Sensor Positioning Uncertainties in the Structural System with Tonal Excitation

The sensitivity of the structure to the tonal input excitation is studied here. The same run-up noise introduced in the acoustic section, Equation 3.35, is used to excite the structural system and the result is plotted in Figure 3.33:

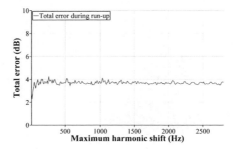

Figure 3.33: Total error due to increase of the first harmonic of tonal excitation in plate

The total error due to tonal excitation shows a relatively constant deviation with shifting frequency. Comparing the relative SD analysis of the plate with broadband and tonal input excitation; the tonal and broadband excitations have the same influence on the uncertainty analysis of the structural system.

3.4 Statistical Analysis of the Transfer Path above Schroeder Frequency

It is seen that above the Schroeder frequency the uncertainties saturate and don't increase with the frequency. Therefore, this is considered as a histogram of the transfer path above the Schroeder frequency due to each uncertainty parameter and the distribution of the samples in this region is studied. First, sensor positioning in the area 1 is considered. As mentioned earlier the samples are normally taken from the reference position of the receiver, the histogram of sensor positions with respect to the sound source position is plotted in Figure 3.34.

Figure 3.34: Histogram of sensor positioning in area 1

A histogram of the sound pressure of sample number 200 above the Schroeder frequency in area 1 of sensor positioning is plotted in Figure 3.35:

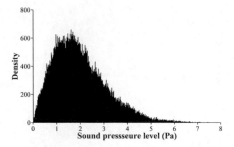

Figure 3.35: Histogram of sound pressure level above the Schroeder frequency for the sample number 200 in area 1

It can be seen that the distribution is similar to the Rayleigh distribution. The histogram of the 400 samples is taken and compared with the reference sample using the Kolmogorov–Smirnov test (K–S test or KS test) which compares any significant differences of two distributions. The K–S test in this analysis returns the h value of one and p-value lower than 0.05, thus the differences between each distribution are significant. The mean and variances of 400 Rayleigh samples are plotted in Figure 3.36:

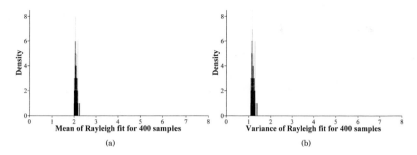

Figure 3.36: Mean (a), variance of Rayleigh fit of 400 samples in area 1 (b)

The K–S test has been applied to all the uncertainty analysis cases to indicate any significant differences of sound pressure in the overlap region.

In the proposed system with potential temperature deviation, the K–S test reveals that the temperature changes of about $10°C$ have a non-significant deviation of sound pressure distribution at higher frequencies with the confidence interval of 95%. The p-values of all histograms obtained in the temperature uncertainty analysis are compared in Table 3.1. The gray color indicates that the deviation between the reference sample and the samples with specified temperature are significant and the rest are insignificant.

Table 3.1: Statistical K–S analysis for the system with temperature deviations

	$T_{15°C}$	$T_{16°C}$	$T_{17°C}$	$T_{18°C}$	$T_{19°C}$	$T_{20°C}$	$T_{21°C}$	$T_{22°C}$	$T_{23°C}$	$T_{24°C}$	$T_{25°C}$	$T_{26°C}$	$T_{27°C}$	$T_{28°C}$	$T_{29°C}$	$T_{30°C}$
$P_{15°C}$	1.0000	0.9985	0.8435	0.8605	0.4729	0.3351	0.1743	0.1527	0.1079	0.0911	0.0175	0.0128	0.0077	0.0014	0.0006	0.0001
$P_{16°C}$	0.9985	1.0000	0.9999	0.9998	0.9061	0.7248	0.4959	0.4882	0.3111	0.2746	0.0933	0.0691	0.0353	0.0086	0.0045	0.0009
$P_{17°C}$	0.8435	0.9999	1.0000	1.0000	1.0000	0.9690	0.8828	0.8294	0.6128	0.6004	0.2538	0.1962	0.1272	0.0334	0.0175	0.0042
$P_{18°C}$	0.8605	0.9998	1.0000	1.0000	1.0000	0.9875	0.8074	0.7807	0.6128	0.5880	0.2389	0.1687	0.1092	0.0277	0.0195	0.0036
$P_{19°C}$	0.4729	0.9061	1.0000	1.0000	1.0000	1.0000	0.9966	0.9690	0.9061	0.8435	0.4805	0.3902	0.2112	0.0825	0.0518	0.0121
$P_{20°C}$	0.3351	0.7248	0.9690	0.9875	1.0000	1.0000	1.0000	1.0000	0.9976	0.9903	0.6586	0.5757	0.4395	0.2025	0.1317	0.0368
$P_{21°C}$	0.1743	0.4959	0.8828	0.8074	0.9966	1.0000	1.0000	1.0000	1.0000	0.9999	0.9167	0.8504	0.6753	0.3701	0.2438	0.0795
$P_{22°C}$	0.1527	0.4882	0.8294	0.7807	0.9690	1.0000	1.0000	1.0000	1.0000	1.0000	0.8702	0.8185	0.7207	0.3701	0.2270	0.0815
$P_{23°C}$	0.1079	0.3111	0.6128	0.6128	0.9061	0.9976	1.0000	1.0000	1.0000	1.0000	0.9360	0.8258	0.8858	0.5353	0.3636	0.1493
$P_{24°C}$	0.0911	0.2746	0.6004	0.5880	0.8435	0.9903	0.9999	1.0000	1.0000	1.0000	0.9676	0.9267	0.8888	0.5554	0.4039	0.1801
$P_{25°C}$	0.0175	0.0933	0.2538	0.2389	0.4805	0.6586	0.9167	0.8702	0.9360	0.9676	1.0000	1.0000	1.0000	0.9757	0.8734	0.4920
$P_{26°C}$	0.0128	0.0691	0.1962	0.1687	0.3902	0.5757	0.8504	0.8185	0.8258	0.9267	1.0000	1.0000	1.0000	0.9890	0.9446	0.6004
$P_{27°C}$	0.0077	0.0353	0.1272	0.1092	0.2112	0.4395	0.6753	0.7207	0.8858	0.8888	1.0000	1.0000	1.0000	0.9997	0.9921	0.7729
$P_{28°C}$	0.0014	0.0086	0.0334	0.0277	0.0825	0.2025	0.3701	0.3701	0.5353	0.5554	0.9757	0.9890	0.9997	1.0000	1.0000	0.9850
$P_{29°C}$	0.0006	0.0045	0.0175	0.0195	0.0518	0.1317	0.2438	0.2270	0.3636	0.4039	0.8734	0.9446	0.9921	1.0000	1.0000	0.9989
$P_{30°C}$	0.0001	0.0009	0.0042	0.0036	0.0121	0.0368	0.0795	0.0815	0.1493	0.1801	0.4920	0.6004	0.7729	0.9850	0.9989	1.0000

In the system with uncertainties in the damp factor, the K–S test indicates the significant deviation for all cases with the p-value far less than 0.05. The mean and variance of the Rayleigh fit are shown in Figure 3.37. Despite the K–S test results, the results of mean and variances are close to each other.

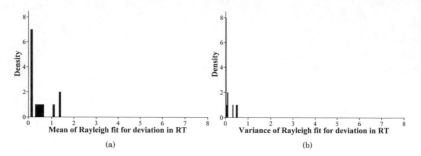

Figure 3.37: Mean (a), variance of Rayleigh fit of 400 samples in area 1 (b)

These analysis results show that in the case of combined analysis with sensor positioning and damping deviation, the dominant factor is damping uncertainties rather than sensor positioning uncertainty.

In the structural analysis considering the overlap frequency above 976 Hz, the distribution of the acceleration level above the overlap frequency is obtained as well. The mean and variance of the Rayleigh distributions of the acceleration level are plotted in Figure 3.38. The K–S test indicates that in the structural analysis deviation of the acceleration level is significant.

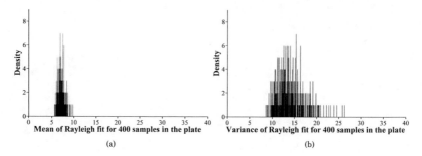

Figure 3.38: Mean (a), variance of Rayleigh fit of 400 samples in area 1 (b)

3.5 Summary and Discussion

In this chapter, firstly the concept of uncertainty analysis is applied to the simple acoustic system with sensor positioning, temperature variation and damping deviation. Then, the same theory is applied to the structural model with sensor positioning uncertainties. The following results are drawn from the first part of the uncertainty modeling:

- The relative SD of the acoustic system with the uncertainty sources increases with the frequency, however, above the overlap frequency, the error saturates. In this case study, the error saturation is around 4 dB for the external sources of uncertainties: sensor positioning and temperature; and for the internal uncertainty source such as changes in the damp factor, the error saturates at 6 dB.

- In the case of sensor positioning uncertainties in the acoustic system, the increase of the sensor positioning error around the reference receiver has no influence on the error saturation.

- Changes in the reference sample position only have an influence on the middle frequency, while the saturation stays the same.

- The relative SD analysis for the sensor positioning uncertainties in the structural system slightly increases and above the overlap frequency the error is flat at 4 dB.

- The influence of the tonal input excitation on the uncertainty analysis is studied in each case. The relative SD of each case of the uncertainty model shows a similar behavior in the same system with the broadband excitation in the diffuse field.

The saturation in the statistical region of each transfer path is studied through the K–S test and the results can be itemized as below:

- The sensor positioning and damp factor deviation show the significant difference of the Rayleigh fit for each pair of the uncertainty model, however the sensor positing uncertainties show very narrow mean and variance deviations.

- The temperature deviation shows the insignificant differences of the Rayleigh fits of the final measured transfer path with temperature deviation of $10°C$. Above these differences the higher frequencies also result in significant differences.

Finally, a statistical approach is applied to study the significant differences of distribution above the overlap region in both systems. The statistical results in this system reveal that the temperature deviation of about $10 °C$ has no significant effect on the final measured transfer path even in the higher frequencies.

4

Experimental Uncertainty Analysis of Transfer Path

A measurement of a system with respect to the possible uncertainty parameters is challenging and requires a clear vision about the purpose of the measurement. The main focus of this chapter is to assess the relative error range in the measurement of the transfer path with respect to the sensor positioning, temperature, and damping deviations. Therefore, this chapter first shows: the procedure of the transfer path measurement with sensor positioning uncertainties in acoustic, structural, and vibro-acoustic systems. Secondly, the uncertainties in the measured signals are assessed according to the uncertainty analysis method. Finally, the transfer path due to each source of uncertainties in the overlap region is statistically studied through the K–S test method.

4.1 Experimental Setup

A typical acoustic measurement chain is given in Figure 4.1 (DIETRICH, 2013). The measurement chain is split into input and output measurement chains. The output chain consists of the DA converter, power amplifier and actuator. This chain is developed to transfer the excitation signal from the software to the actuator. The actuator excites the system under test and the input chain is arranged to carry the response of the system to the software through the sensor, power pre-amplifier and AD converter. A flat frequency response of the measurement chain is required to perfectly measure the transfer path of the device under test (DUT). The following gives an explanation of the sensitivity of each element in the measurement chain to create a flat frequency response.

> *AD converter:* this element quantifies the measured signal and has a nonlinear frequency response. Depending on the quantization level, the nonlinear behavior of the element can be increased; however, a linear frequency response of the AD converter is assumed in the frequency range of interest.

DA converter: this element also has a nonlinear behavior; however, it can be ignored in this measurement.

Power amplifier and pre-amplifier: they show a linear behavior with a flat frequency response. Theses devices are designed to suppress noise as well.

Sensor: in the case of the omni-directional microphone, they have a relatively flat frequency response in the audible frequency range.

Actuator: this device is very frequency dependent, thus the frequency response of the actuator should be suppressed after the measurement. (DIETRICH, 2013)

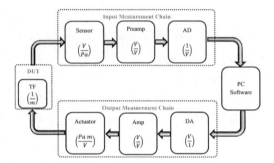

Figure 4.1: Measurement chain DIETRICH, 2013

In the measurement of the acoustic system, the sensor and actuator are omni-directional microphones and loudspeakers, whereas in the measurement of the structural system, sensor and actuator are shakers and accelerometers. A list of the equipment is provided in the appendix. In order to assure the flat frequency response of the measurement chain, calibration is required of each element of the measurement chain. All the experiments in this thesis are relative measurements, thus an absolute calibrated measurement chain is not privileged, nevertheless for the sake of completeness the measurement chain is calibrated.

Each element of the measurement chain is a source of uncertainties. The uncertainties in the input and output chain are comprehensively discussed by DIETRICH, 2013. In this thesis, the measurement chain is unchanged during the relative study in all the cases and the uncertainties in the system under test are investigated. Uncertainties can be introduced due to the imperfection placement of the sensors and actuator, or the environmental changes like deviation in the temperature, and internal changes of the DUT, which lead to variation of the reverberation or decay time (damp factor variation).

4.2 Acoustic Transfer Path Measurement

The geometry being tested is a rectangular enclosure, made of MDF with dimensions of $0.8 \times 0.5 \times 0.3 \ m^3$ and thickness of 22 mm, Figure 4.2(a). The top side of the enclosure is removable and the exterior edges of the walls are sealed with a foam strip to eliminate the air gap. The plate on the top side of the enclosure is attached to the box with eight sealing clamps to compress the foam strip (MOHAMADY et al., 2013).

An omni-directional loudspeaker is placed in the corner of the enclosure and a x-y table is embedded in the bottom of the box. The x-y table is designed to carry four microphones that are capable of moving in both directions (x-y) with resolution of 1 mm, Figure 4.2(b). An interface is developed to control the x-y table, while the enclosure is completely sealed. The environmental changes during the measurement are monitored using three temperature and humidity sensors inside the box. The connection wires are passed through two small opening areas in the wall of the enclosure; likewise, an air gap sealing is made for these openings.

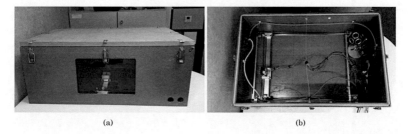

(a)   (b)

Figure 4.2: Exterior view of the rectangular enclosure made of MDF (a), interior design of the enclosure (b)

4.2.1 Input Excitation

Two types of input excitation sources are applied to the system: broadband and tonal excitation. A sweep signal according to MÜLLER and MASSARANI, 2001 is used as the broadband excitation to actuate the omni-directional loudspeaker through the measurement chain. The spectrogram of the sweep signal is illustrated in Figure 4.3.

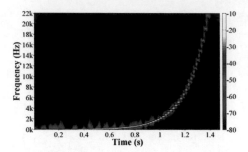

Figure 4.3: Spectrogram of the exponential sweep signal

The tonal excitation is generated using a scaled electric engine. The electric engine is a permanent magnet synchronous machine (PMSM), series 19, model 1920, Figure 4.4. It represents a scale model of an electric engine mounted in a smart electric drive. The engine noise is measured during the run-up and at specific speeds.

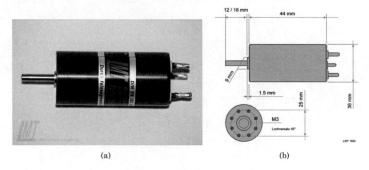

(a)   (b)

Figure 4.4: Electric engine PMSM (a), and details (b)

The engine is mounted inside the enclosure with two scrolls, Figure 4.5.

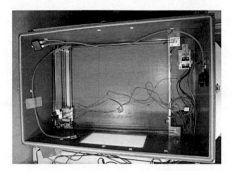

Figure 4.5: Transfer path measurement of 400 samples

A sample of measured engine noise during the run-up inside the enclosure is shown in Figure 4.6.

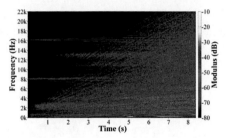

Figure 4.6: Engine run-up measurement inside the enclosure

The influence of the transfer path of the enclosure on the measured engine noise during the run-up can be captured from Figure 4.6. Moreover, the reluctance noise at 8 and 16 kHz can be observed.

4.2.2 Uncertainty Analysis

Three sources of uncertainties are introduced in the current measurement: the sensor displacement, temperature changes and variation in damping (decay time). The procedure of developing the measurement setup for the system with these uncertainties is explained below:

Sensor Positioning Uncertainties

The sensor positioning measurement is developed to be identical with the analytical model. In this regard, an x-y table is designed to scan the area of interest inside the enclosure, Figure 4.7.

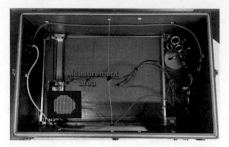

Figure 4.7: Transfer path measurement of 400 samples

Four microphones are embedded on the moving part of the x-y table upward in the z direction, the resolution of each step is 1 mm, and with each movement the transfer path is measured inside the enclosure. A total of 7744 samples are taken from the area 1 with a measurement surface of $10 \times 10 \ cm^2$, see the red surface in Figure 4.8. After taking the measurement, 400 samples are chosen normally from the middle part of the measurement area (light dots inside the measurement area). The transfer path between the excitation source and chosen 400 receivers is plotted in Figure 4.8:

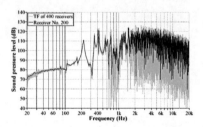

Figure 4.8: Transfer path measurement of 400 samples

The dark color plot in this figure indicates the reference transfer path and the light color shows 400 samples. The relative standard deviation of 400 samples is calculated according to equation 3.31 and 3.32 and plotted in Figure 4.9.

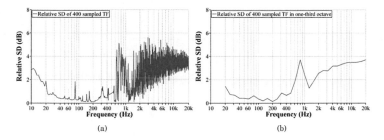

(a)   (b)

Figure 4.9: Relative SD analysis of measured 400 TF (a), averaged power of the relative uncertainties analysis in one-third octave bands (a)

The results in the one-third octave band show an increase of error in the frequency and saturation above the Schroeder frequency. The high error in the lower frequencies is due to the measurement noise. Along the increase of the error, a high peak around 1 kHz is observed, which could be due to the specific reference sampling position. The sensor positioning measurement is made in the four individual areas inside the box, with the same strategy explained in the analytical chapter in subsection 3.3.1. The scanned area in the experimental setup is shown in Figure 4.10.

Figure 4.10: Experimental setup to measure area 1 to area 4

The results of the relative uncertainty analysis of all areas are shown in Figure 3.2.1.

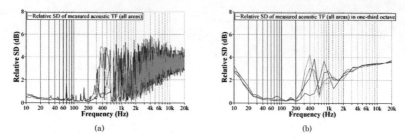

Figure 4.11: Relative SD analysis of measured 400 TF (a), averaged power of the relative uncertainties analysis in one-third octave bands (b)

The result of the analysis in four areas inside the box highlights the dependencies of the error propagation to the reference sample position in the middle frequencies and the independences to the reference sample position in the higher frequencies (error saturation). These results will be used to evaluate the uncertainty modeling in the next chapter.

Temperature Uncertainties

Temperature changes in the laboratory condition are demanding. In this experiment, temperature is deliberately changed from $20°C$ to $27°C$ externally. The transfer path is measured in a position inside the enclosure during the increase of the temperature. Three sensors are attached to the interior side of the enclosure to measure the temperature and humidity changes during the measurement, Figure 4.2(b). The deviation of temperature and humidity versus time is plotted in Figure 4.12:

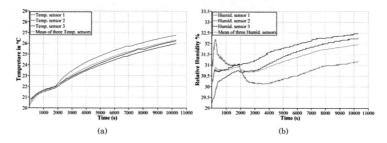

Figure 4.12: Temperature (a) and humidity (b) variation in time

During the temperature changes, in total, 700 transfer paths are measured and plotted in Figure 4.16:

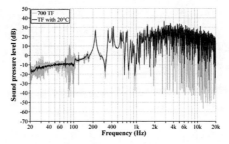

Figure 4.13: Transfer path measurement of 700 samples with temperature deviation

The dark color shows the transfer path with the lowest temperature and the light cloud indicates the 700 transfer paths of the system. The samples were recorded in two hours and 51 minutes with a maximum interior temperature of $26.3°C$. The uncertainty analysis according to equation 3.31 and 3.32 is calculated and presented in Figure 4.14.

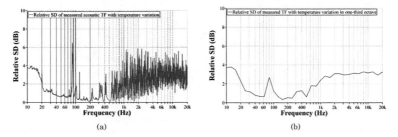

(a)   (b)

Figure 4.14: Relative SD analysis of measured 700 TF (a), averaged power of the relative uncertainties analysis in one-third octave bands (b)

A relatively uniform increase of error in frequency can be observed during the measurement; however, a high peak is detected at 100 Hz which can be interpreted as the measurement sampling position. The saturation of error deviation above the Schroeder frequency is detectable as well.

Damping Uncertainties

The damping uncertainties are measured with the variation of the absorption material inside the box. Three types of the absorption material are combined and placed inside the enclosure with 17 configurations and the transfer paths are measured in a reference point via an omni-directional loudspeaker. A number of measurement cases are presented in Figure 4.15.

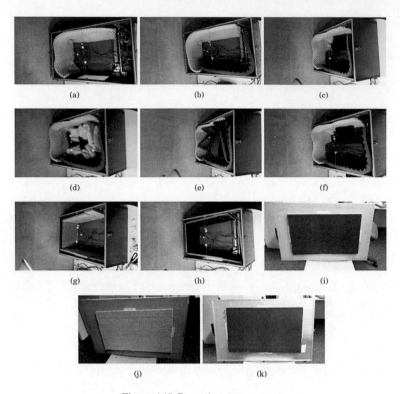

Figure 4.15: Decay time measurement

The impulse responses are measured at four positions inside the enclosure and the decay time is calculated according to ISO 3382-1, 2009. The transfer path of the system with 17 cases of absorption materials is illustrated in Figure 4.16. The computed decay times in 17 cases are plotted in Figure 4.17:

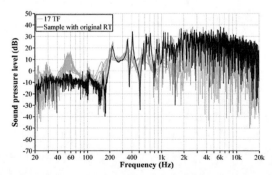

Figure 4.16: Measure 17 transfer paths with a variety of absorption material

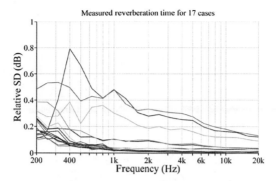

Figure 4.17: Decay time for 17 cases

In this graph, the enclosure without any absorption material is plotted in a dark color and the rest in light color. The relative SD is calculated and illustrated in Figure 4.18. The results show the saturation of error around 6 dB above the Schroeder frequency.

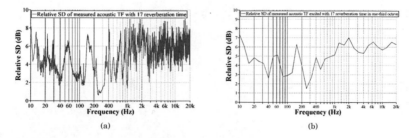

Figure 4.18: Relative SD analysis of the measured 17 transfer paths (a), averaged power of the relative SD analysis in one-third octave bands (b)

The fluctuation of the error analysis is due to the shape of the observer material, however the tendency shows the error increase and saturation as well.

4.3 Structural Path Measurement

The setup to measure the structural transfer path is shown in Figure 4.19. The system is a rectangular aluminum plate with the dimensions $0.8 \times 0.5\ m^2$ and thickness of $1\ mm$. The excitation source is a shaker, which is placed under the plate, and vibrates the plate from the down side. The accelerometer is placed on the top side of the plate which measures the acceleration created by the shaker. The transfer path of the plate is measured using the second accelerometer at a specific distance from the source position. In total, the transfer path is measured in five positions.

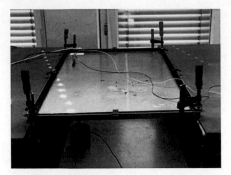

Figure 4.19: Measurement setup of the aluminum plate

The shaker has the excitation frequency limit up to $4\ kHz$, thus, the measured acceleration above this frequency is invalid. The five measured transfer paths are plotted in Figure 4.20 and their relative SD analysis with sensor positioning uncertainties is shown in Figure 4.21. Below the overlap frequency, $976\ Hz$, the relative SD is fluctuating around 4 dB and above this frequency, constant behavior can be observed.

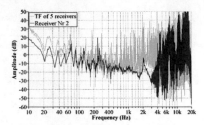

Figure 4.20: Transfer path measurement in five different receiver positions

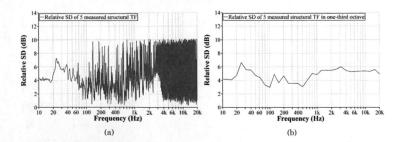

Figure 4.21: Relative SD analysis of measured 5 TF (a), averaged power of the relative uncertainties analysis in one-third octave bands (b)

According to Figure 4.21, the structural system is very sensitive to the sensor positioning uncertainties even in the modal region.

4.4 Vibro-Acoustic Transfer Path Measurements

The vibro-acoustic transfer path of the system is arranged by placing an aluminum plate on top of the enclosure. The measurement is made in two phases, in the first phase, the airborne transfer path is measured inside the box with the aluminum plate attached to the box, Figure 4.22(a). In the second phase, the structural transfer path is measured with a VISATON shaker as the excitation source on top of the aluminum plate and an accelerometer is placed in the other corner of the plate, Figure 4.22(b).

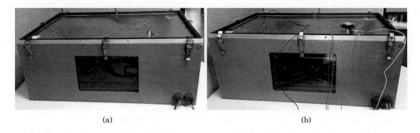

(a)   (b)

Figure 4.22: Rectangular enclosure with an aluminum plate on top side in phase 1 (a), phase 2 (b)

In the first phase, 400 samples are taken around the reference receiver position (area 1 in the acoustic TF measurement), and in the second phase 78 samples are measured on the top of the plate. The transfer paths measured in both phases are plotted in Figure 4.23.

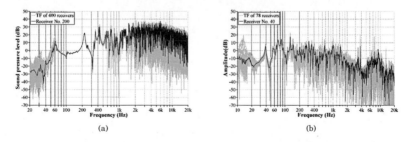

(a)   (b)

Figure 4.23: Vibro-acoustic measurement (phase 1) (a), vibro-acoustic measurement (phase 2) (b)

The first mode of the enclosure can be captured from both analyses at 232 Hz. The third mode of the plate also contributes to the transfer path measurement of both systems. The relative SD of two setups is calculated to evaluate the saturation in the coupled vibro-acoustic systems, Figure 4.24.

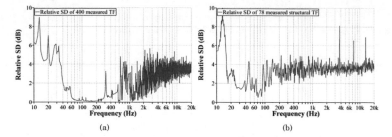

(a)   (b)

Figure 4.24: Relative SD analysis of vibro-acoustic system measurement (phase 1) (a), relative SD analysis of vibro-acoustic system measurement (phase 2), (b)

The averaged power of the relative SD of both systems in one-third octave band is calculated and shown in Figure 4.25. The error analysis in both cases shows that the saturation of error is around 4 dB. In the first phase, the saturation is above the overlap frequency of the enclosure; however in the second phase, the saturation of error is reduced and starts from the overlap frequency of the plate.

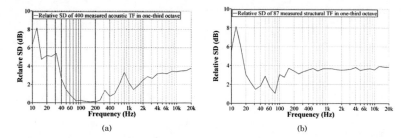

(a)   (b)

Figure 4.25: Averaged power of the relative uncertainties analysis in one-third octave band (phase 1) (a), averaged power of the relative uncertainties analysis in one-third octave band (phase 2) (b)

4.5 Statistical Analysis of the Transfer Path above the Schroeder Frequency

The observed error saturation at the higher frequencies is statistically analyzed in this section. In the first step, the effect of sensor positioning is considering uncertainty in the acoustic measurement of area 1. A histogram of the sensor positions in area 1 is plotted in Figure 4.26.

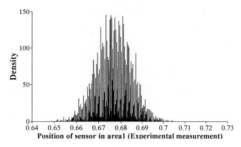

Figure 4.26: Histogram of sensor positioning in area 1

The histogram of the sound pressure level above the Schroeder frequency of the sample number 200 is plotted in Figure 4.26:

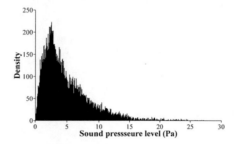

Figure 4.27: Histogram of the sample number 200 in area 1

The histogram of all 400 samples is obtained and the Rayleigh fit of them is calculated. Furthermore, the mean and variances of the Rayleigh distributions of sound pressure are compared and shown in Figure 4.28:

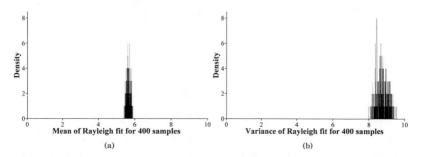

Figure 4.28: Mean (a) Variance (b) of Rayleigh fit of 400 samples in area 1 (Experimental setup)

The comparisons reveal that the mean variations are very small, although the variances show a wider deviation. The 400 distributions are compared via K–S test to check the significant differences between the 400 sampled Rayleigh distributions. The analyses indicate the p-value less than 0.05 which shows the significant differences in 95%.

The same procedure of statistical analysis is performed on the results of the uncertainty analysis with the temperature variation. The mean and variance of the Rayleigh fit of 700 transfer paths above the Schroeder frequency are plotted in Figure 4.29. The detected variation in the mean value of the Rayleigh fit is very small, but variances show higher deviation. Despite the small deviation of mean values, the K–S tests show significant differences in the Rayleigh distributions.

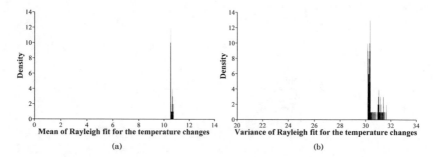

(a)   (b)

Figure 4.29: Mean (a) and Variance (b) of Rayleigh fit for 400 samples in area 1 (Experimental setup)

The Rayleigh fit of the transfer paths with changes in the absorption material is observed and the means and variances are plotted in Figure 4.30. It can be easily observed that the Rayleigh distributions are not similar and the propagation of the uncertainties above the Schroeder frequency are dependent on the absorption materials.

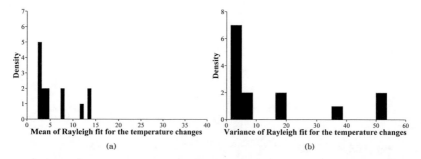

(a)   (b)

Figure 4.30: Mean (a) and Variance (b) of Rayleigh fit for 400 samples in area 1 (Experimental setup)

4.6 Summary and Discussion

In this chapter an experimental setup is presented to measure the relative uncertainties in the acoustic and structural system. In the acoustic system, the uncertainties in sensor positioning, temperature and damping are assessed and in the structural system, the influence of sensor positioning uncertainties is measured. The following results are obtained from the analyses:

- The measurement of the acoustic system with the sensor positioning uncertainties shows an increase of the error in comparison with the reference sample and above the Schroeder frequency saturation of error can be captured.

- The uncertainty analysis in sensor positioning is performed for four reference sensor positions. The results indicate the dependencies of the middle frequencies to the sensor positioning and independences to the saturation of error in the higher frequencies.

- The transfer path measurement with uncertainties in the temperature is also performed and the results indicate the increase of error in the modal region of the transfer path and saturation at 4 dB in the overlap region of the transfer path.

- The setup to measure the transfer path with 17 absorption materials was arranged and the transfer paths were compared, the relative SD analysis shows the increase of error and saturation at 6 dB in the frequency domain.

Two setups are arranged to measure the acoustic and structural transfer paths in a complex vibro-acoustic system. The sensor positioning uncertainties were measured in both cases and the relative SD analysis indicates that in the acoustic transfer path measurement the saturation shifted toward the overlap region, while in the structural transfer path measurement of the vibro-acoustic system the relative SD saturates error earlier than the overlap region of the plate. It can be concluded that with an increase in the complexity of the vibro-acoustic system, the error saturation occurs earlier in the frequency domain, and the error can be assigned as a single value.

In general, the measurement results confirm the results from modeling in Chapter 3, which was the main purpose of implementing the measurement setups.

As a last step, the statistical assessment is performed to observe the significant differences of the error propagation in all uncertainty measurements at higher frequencies. The K–S tests show that all the uncertainties have significant differences; however, the width of Rayleigh fits in the case of damping uncertainties is very large in comparison with the sensor positioning and temperature changes. Thus it is concluded that the internal changes of the structure during measurement have a greater influence in comparison with the external sources.

5

Evaluation and Generalization of Uncertainty Analysis

This chapter can best be treated under two headings: Verification and generalization of the uncertainty analysis. The purpose of the verification analysis is to evaluate the analytical uncertainty model. In this regard, the results of the uncertainty analysis in the analytical part are compared with the experimental one. In the second part, after verification of the results the uncertainty analysis is expanded for general cases.

5.1 Verification of Uncertainty Analysis

The uncertainty model corresponding to each source of uncertainties in the acoustic and structural systems is verified using the experimental error analysis.

5.1.1 Verification of the Acoustic System

Three sources of uncertainties are considered in the acoustic system: Sensor positioning, temperature changes and damping. All these sources of uncertainties are analytically calculated, and the experimental arrangements are performed to verify the analytical uncertainty models. The averaged power of the relative SD in one-third octave band with sensor positioning uncertainties in area 1 is compared for both analytical and experimental cases, and the results are illustrated in Figure 5.1:

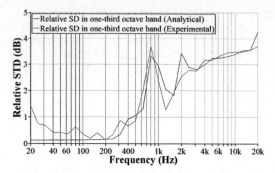

Figure 5.1: Comparing uncertainty model of acoustic system

The comparison of the analytical and experimental uncertainty analysis shows that very good agreement is achieved. In the lower frequencies, the deviation in the experimental results is due to the measurement noise.

The uncertainties in the temperature variations are also verified by comparing the averaged power of the relative SD in one-third octave band. The analysis in the middle frequency range is similar; however, the analytical model shows higher error due to temperature changes. In both analyses, the patterns of the error propagation through frequency are the same.

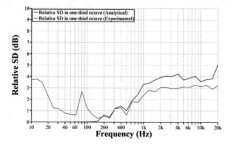

Figure 5.2: Comparing the averaged power of the relative SD in sensor positioning in area 1 of the acoustic system

The relative SD of the error propagation due to a variety of decay times is compared in Figure 5.3.

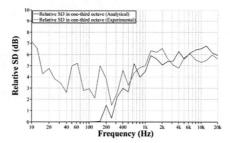

Figure 5.3: Comparing uncertainty model of acoustical system

In the middle and high frequencies, the averaged power of relative SD shows very good agreement, the higher error amplitude in the lower frequencies is due to measurement noise as well.

5.1.2 Verification of the Structural System

The uncertainties in the structural analysis are verified by comparing the error propagation due to the sensor positioning uncertainties in the analytical and experimental studies, Figure 5.4.

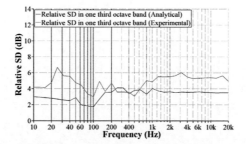

Figure 5.4: Comparing uncertainty model of structural system

The relative SD analysis shows that the structural error fluctuates about 4dB; however, above the overlap frequency, the deviation of error analysis is flat at 4 dB. The experimental and analytical structural uncertainty analysis shows $2\ dB$ differences in the overlap region.

5.2 Generalization of Uncertainty Analysis

In this section, the possible rescaling method is proposed to generalize the uncertainty analysis for the practical cases. In the process of calculating the relative SD, the amplitude of the transfer path is canceled out, therefore, the generalization gives the same relative uncertainty deviation, while the frequency domain should be adjusted to the interested system. Two parameters should be considered in the rescaling of the frequency domain; the geometrical length of the system and wavelength. By describing the volume of the geometry with V, the geometrical length calculates as $L_g = (V)^{1/3}$ and wavelength with $\lambda = \frac{c}{f}$. Thus, the uncertainty analysis could be presented by the following normalized frequency (rescaling factor):

$$\frac{L_g}{c} f = \frac{L_g}{\lambda} \tag{5.1}$$

The results of the sensor positioning uncertainties in the acoustic system (Figure 3.17) are rescaled as below:

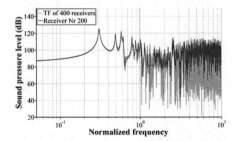

Figure 5.5: Transfer path of the enclosure with 400 receivers

The relative SD of sensor positioning in four areas (chapter 3) is reported with normalized frequency in Figure 5.6.

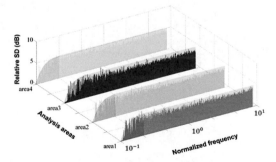

Figure 5.6: Relative SD analysis of area 1 to 4 in the normalized frequency scale

The averaged power in the same area is calculated and reported in the general case in Figure 5.7:

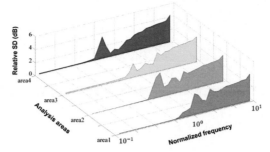

Figure 5.7: Averaged power of relative SD in area 1 to 4 in the normalized frequency scale

The maximum error of 4 dB is expected for the relative maximum sensor placement of $(r/L_g) * 100 = 6\%$ with respect to the geometry length of the system. The width of the error with the sensor positioning uncertainty is independent of the reference excitation position in the statistical region; while the error propagation in the mid-frequency region is influenced by the reference sensor position (variation of peak error before the error saturation).

The generalization is applied to all sources of uncertainties in the acoustic system. The averaged power analysis with three uncertainty sources is plotted together with the normalized frequency in Figure 5.8.

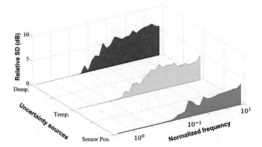

Figure 5.8: Transfer path of the enclosure with 400 receivers

The temperature changes in the range of $\Delta T = 15°C$ introduce maximum deviation of 4 dB. The decay time is frequency dependent and could be a source of high error in the final analysis. Obviously, it can be inferred from the results that the external sources of uncertainties like sensor positioning and temperature variation introduce less error in the measurement process than internal changes within the system. Furthermore, the structural transfer path analysis could be generalized with the same method.

5.3 Summary and Discussion

The chapter is mainly developed to evaluate the uncertainty analysis in the analytical section. According to the comparison results, the evaluation is valid for all the uncertainty cases. Following the consistent agreement of the uncertainty model, the system is generalized to the practical examples by introducing the scale factor to normalize the frequency range of interest. The following results are obtained from the generalization of error analysis:

- The uncertainty in sensor placement with a distance of 6% of the geometrical length of the system is about 4 dB.

- The temperature changes up to $15°C$ introduce the maximum error of 4 dB.

6

Consequences of Uncertainties for Sound Design

Sound design involves the acquisition and manipulation of a sonic event to form a desired sound in a specified scenario which is subject to objective and subjective analysis. The objective analysis focuses on the physical phenomenon, while the subjective analysis considers human perception and could be analyzed through the psychoacoustic approach and listening test. The question to be considered here is how the input uncertainty changes, such as variation in the sensor positioning, may have an influence on the sound design parameters. This question can be answered by the following analytical steps:

Uncertainty modeling: analyzing the uncertainties of the system with a variety of input excitations.

Post-processing: obtaining the psychoacoustic parameters and performing the listening test as post-processing of the uncertainty model toward sound design of the system under test.

Uncertainty analysis: linking the deviation analysis of the objective and subjective outputs from the post-processing step to the uncertainty model.

This approach is shown as a block diagram in Figure 6.1.

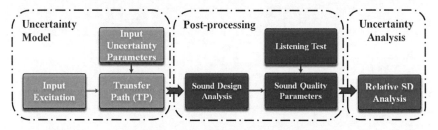

Figure 6.1: Sound design uncertainty analysis

According to Figure 6.1, the uncertainty model consists of a transfer path of the vibro-acoustic system with the input excitation and input uncertainty parameters. The uncertainty modeling block provides a set of the transfer paths with each uncertainty parameter. The post-processing block determines the influence of the input uncertainties on the psychoacoustic parameters and indicates the sensitivity of human hearing to the uncertainties. The uncertainties are calculated in the post-processing step, and finally, the range of the deviation in the uncertainty analysis block is compared with the input uncertainty parameters. These analysis processes are explained explicitly in the next subsections.

6.1 Case Study - Experimental Example

The uncertainty model describes the behavior of the system with defined input excitations and feasible uncertainties. A simple acoustic system with four types of input excitations: Engine noise at low, middle and high speed and pink noise is defined as a system under test. An input uncertainty parameter is variation in the position of the measurement sensor.

6.1.1 Sample Transfer Path

The acoustic system introduced in chapter 2 is used as a sample of the transfer path in the uncertainty model in Figure 6.2. Its dimensions are $0.8 * 0.5 * 0.3\ m^3$ with interior sound source and receivers. The sound source position is shown with an omni-directional loudspeaker symbol and the receivers with a thick line in the upper left side of the box. The reference receiver position is denoted with a bold point in the top far left-hand corner of the receiver line. The transfer path between the sound source and reference receiver is plotted in Figure 6.3. The first eigenfrequency of the system is placed at $232\ Hz$ and the Schroeder frequency at $3.6\ kHz$.

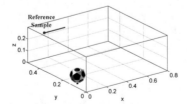

Figure 6.2: Sound design uncertainty analysis

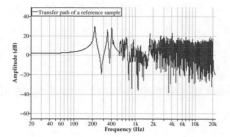

Figure 6.3: Transfer path of reference sample

6.1.2 Input Excitation

The input excitations are grouped in two categories of tonal and broadband excitation. Electric engine noise is used as a sample of tonal noise and pink noise as a broadband excitation. The electric engine used in this study is the permanent magnet synchronous machine (PMSM), series 19, model 1920, Figure 4.4. The electric engine noise is measured according to the input measurement chain shown in Figure 6.4.

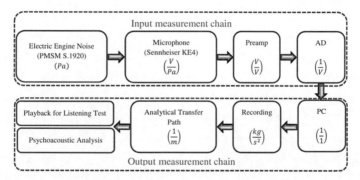

Figure 6.4: Electric engine noise measurement chain

The electric engine noise is measured while mounted on a hard surface to block the boundary condition at the contact points. The electric engine noise and vibration are measured using a microphone and laser vibrometer. The measurement setup is shown in Figure 6.5. All signals are measured with a resolution of 24 bit and sampling frequency of 44.1 kHz.

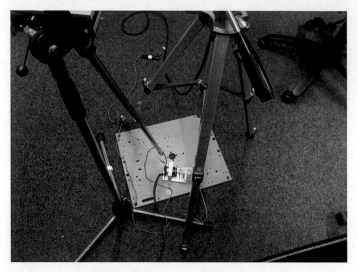

Figure 6.5: Electric engine noise measurement

Three operational speeds of the electric engine namely at low (5590 rpm), middle (10510 rpm) and high (14200 rpm) speed are measured. The fundamental frequency of the electric engine noise at low, middle and high speed are $48Hz$, $179Hz$ and $248Hz$, respectively. The spectrogram of the engine noise namely at low, middle and high speeds is plotted in Figure 6.6. A filtered pink noise is simulated as a broadband input excitation and its spectrogram is shown in Figure 6.5. The sound pressure level deviation is shown with colors in all the spectrograms. The SPL increases with frequency at higher engine speed. These four types of input excitations are multiplied with the transfer paths of the system with uncertainties in sensor positioning, (output chain in Figure 6.5).

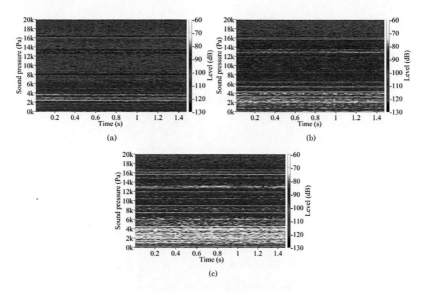

(a)   (b)

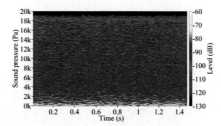

(c)

Figure 6.6: The spectrogram of electric engine noise at (a) low, (b) middle and (c) high speed

Figure 6.7: Spectrogram of pink noise excitation

6.1.3 Input Uncertainty Parameter

A broad range of input uncertainty parameters could be realized during the transfer path measurement, among which, sensor positioning is one of the most expected uncertainty sources. In this study the sensor positioning in the measurement line inside the enclosure is simulated as the input uncertainty. The line displacement facilitates the perception analysis in the feature study. In total 211 transfer paths are calculated within the receiver line inside the box with displacement resolution of 1 mm from the reference point. In Figure 6.8, the transfer path of 211 samples is plotted, the bold color indicates the TF of the reference sample.

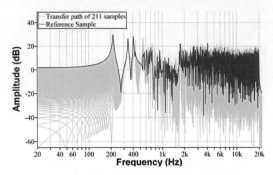

Figure 6.8: Sound design uncertainty analysis

The uncertainty model will be completed by considering the input excitations. There are four groups of input uncertainties consisting of three measured engine noises as tonal input excitations and a pink noise as broadband excitation. The results of the uncertainty model with each group of responses are illustrated in Figure 6.9. The level of stimuli for all four cases is calibrated with 70 dB. In these graphs a transfer path is included of the playback system. The playback transfer path will be explained in the next subsection.

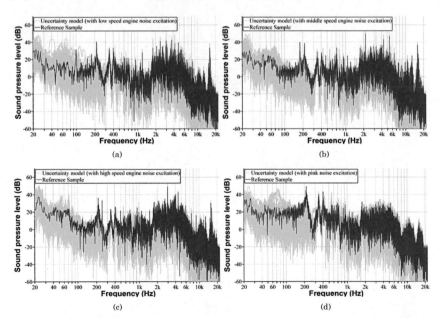

Figure 6.9: Frequency response of the system excited with (a) low speed, (b) middle speed, (c) high speed and (c) pink noise

6.1.4 Transfer Path of the Playback System

The interface between the uncertainty model and the post-processing part in Figure 6.1 is called the playback system. This system provides the signal to be perceived by listeners in the listening test and concurrently used for the psychoacoustic analysis. In Figure 6.10, the block diagram of the playback system is given. The input of this block diagram is the measured engine noise and pink noise convolved with the impulse responses of the system which was explained in subsection 6.1.2.

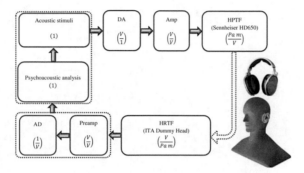

Figure 6.10: Block diagram of playback system

The playback system is configured using a Sennheiser headphone model HD 650 and an ITA Dummy head, plus an audio interface is Focusrite Scarlett 2i2 USB. The measurement chain in Figure 6.10 introduces new changes to the transfer path of the system in the first block. As an example, the transfer path of the reference point inside the enclosure with the low speed engine noise as the input excitation is played via the measurement chain in Figure 6.10 and the result is measured and shown in Figure 6.11. This graph can be compared with Figure 6.9(a) to realize the effect of the playback system.

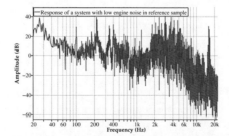

Figure 6.11: The reference transfer path with low engine noise excitation, played via playback system

6.1.5 Uncertainty Analysis

The uncertainty analysis of the system for each input excitation is given below. The analysis is performed according to Equations 3.31, Figure 6.12.

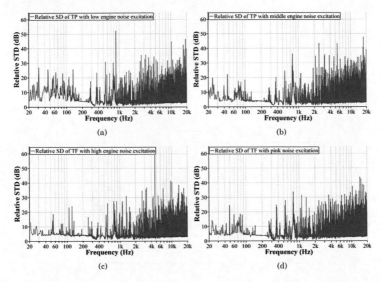

Figure 6.12: Relative SD of system excited with (a) low speed, (b) middle speed, (c) high speed and (d) pink noise excitation

As can be seen in Figure 6.12, when pink noise excitation is applied, random high peaks are less than for the system with engine noise excitations. The results of the uncertainty analysis are calculated according to the averaged power in one-third octave band (Equation 3.32) and plotted in Figure 6.13.

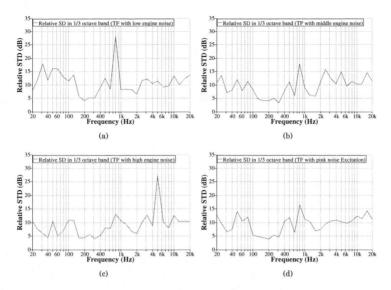

Figure 6.13: Relative SD in 1/3 octave band of system excited with (a) low speed, (b) middle speed, (c) high speed and (d) pink noise excitation

The uncertainty analysis of the system in one-third octave bands indicates the mean variation of 5.6, 8.9, 9.3 and 7.2 dB for the system excited with low, middle, high speed and pink noise excitation. The following interpretations can be drawn from the uncertainty analysis of the system before considering the post-processing step:

1. The systems excited with middle and high speed show larger uncertainties in comparison with the system excited with low and pink noise.

2. The lowest deviation is obtained with the low engine noise excitation, which can be linked to the lower fundamental frequency produced with the electric engine at the low speed, and attributes less amplitude at the high frequencies.

3. The uncertainty deviations in all four conditions are mainly above 5 dB which indicate high sensitivity of the system to the sensor displacement.

The uncertainty analysis in this step will be related to the listening test result which is given at the end of this chapter.

6.2 Post-processing for Sound Quality Parameters

The post-processing block in Figure 6.1 indicates the influence of the uncertainty model introduced in section 6.1 on the sound design analysis by considering the perceptual effects.
Psychoacoustic analysis and the listening test are the sound design tools to evaluate the sound
pattern inside the interested medium. In this regard, firstly a general concept is given of
psychoacoustic analysis including sound quality metrics, then the influence of uncertainties on
these parameters is investigated. Secondly, a listening test is developed to study the influence of
sensor positioning uncertainties on the perception of sound.

6.2.1 Psychoacoustic Analysis

Psychoacoustics is a scientific field to detect hearing sensation. It is developed to define a
quantitative relation between acoustic stimuli (sound) and hearing sensation (impression)
(FASTL and ZWICKER, 2007). Thus understanding the feature of the hearing organ is an essential
step in the psychoacoustic analysis. Psychoacoustic concept rates according to 17 parameters[1]
(PEDERSEN, 2008). The purpose of this chapter is to look into the influence of input uncertainties
on the most important psychoacoustics parameters, e.g: loudness, sharpness, fluctuation strength
and roughness. In the following subsections the concept of the level analysis is explained then
each psychoacoustic parameter is briefly explained and the possible noticeable differences are
discussed. Finally, the variations in these parameters due to the sensor positioning uncertainties
are evaluated.

Loudness Concept

Loudness as an ordered scale from quiet to loud attributes human auditory sensation in the
category of intensity. Standard loudness is defined as a 1-kHz tone at the level of 40 dB which is
named as one sone. To simulate the human hearing spectral perception, two auditory frequency
scales are introduced: Bark and Equivalent Rectangular Band (ERB), both scales are the logarith-
mic function of the frequency (FASTL and ZWICKER, 2007; PULKKI and KARJALAINEN, 2015).
A bandwidth of Bark scale, $\Delta_{f_{Bark}}$ is obtained as:

$$\Delta f_{Bark} = 25 + 75 \left[1 + 1.4 \left(\frac{f_c}{1000} \right)^2 \right]^{0.69} \tag{6.1}$$

---

[1] Loudness, sharpness, roughness, tone prominence, pitch, pitch strength, polyphony, harmony, frequency variation,
localized in space, regularities, tempo, presence, amplitude variation, impulse prominence, duration, decay.

with f_c as a center frequency, the Bark scale can be estimated as (PULKKI and KAR-JALAINEN, 2015):

$$z_{Bark} = 13 \arctan\left(\frac{0.76f}{1000}\right) + 3.5 \arctan\left(\frac{f}{7500}\right)^2 \tag{6.2}$$

A bandwidth of the ERB calculates:

$$\Delta f_{ERB} = 24.7 + 0.108 f_c \tag{6.3}$$

The ERB scale can be estimated:

$$z_{Bark} = 21.3 \log\left(1 + \frac{f_c}{228.7}\right) \tag{6.4}$$

The loudness analysis is computed in the Bark scale. The center frequency, cut-off frequency and bandwidth of the Bark scale are given in Table 6.1:

Table 6.1: Bark scale

Number	Center Frequency (Hz)	Cut-off Frequency (Hz)	Bandwidth (Hz)
1	60	100	80
2	150	200	100
3	250	300	100
4	350	400	100
5	450	510	110
6	570	630	120
7	700	770	140
8	840	920	150
9	1000	1080	160
10	1170	1270	190
11	1370	1480	210
12	1600	1720	240
13	1850	2000	280
14	2150	2320	320
15	2500	2700	380
16	2900	3150	450
17	3400	3700	550
18	4000	4400	700
19	4800	5300	900
20	5800	6400	1100
21	7000	7700	1300
22	8500	9500	1800
23	10500	12000	2500
24	13500	15500	3500

The loudness of pure tone in sone calculates as:

$$N = 2^{\frac{(L_L - 40)}{10}}$$ (6.5)

where L_L is the loudness level of pure tone at 1kHz. Loudness of a broadband excitation is calculated:

$$N = \int_0^M N'(z)dz$$ (6.6)

where $N'(z)$ indicates the specific loudness and M is the number of critical bands. The specific loudness calculates as:

$$N'(z) = CE(z)^{0.23}$$ (6.7)

in this equation, $E(z)$ indicates the excitation pattern and C is a constant to set 1kHz sinusoidal signal at 40 dB to 1 sone. The excitation pattern is calculated as:

$$E(z) = S'(z) \times B(z)$$ (6.8)

where $S'(z)$ is a power spectral density of the sound stimuli entering the hearing mechanism in the auditory scale, and $B(z)$ is a spreading function depending on the level and frequency. In Figure 6.14 a chart of the human hearing organ is provided with the relevant block to calculate the loudness of the broadband signal.

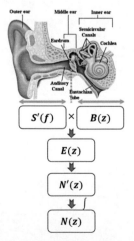

Figure 6.14: Process of calculating loudness

In Figure 6.14, the process of loudness calculation is visually shown. The effective areas in calculation of $S'(f)$ and $B(z)$ are determined with two green arrows. The excitation pattern is obtained by the multiplication of these two functions, and the specific loudness can be obtained with a constant to relate it to the reference loudness as 1 sone. The total loudness is obtained by integration over the frequency range of human hearing.

The value of loudness exceeded in 5% of the measurement time is called N5 and corresponds to the overall perceived loudness (FASTL and ZWICKER, 2007). A just-noticeable difference in loudness is discussed in the next subsections.

Just-Noticeable Differences in Loudness

A subjective listening test was carried out by PEDRIELLI et al., 2008 to study just-noticeable differences (JND) in loudness. A stimulus was recorded at the operator station of an earth moving machine. A reference signal is defined with three loudness levels at 60, 70 and 80 dB. The loudness of each reference signal was varied in a range of ±2.7 sone with a step size of 0.3. The loudness of 32.1, 18.0 and 10.3 *sone* is reported with the sound pressure level of 64.9, 73.1 and 82.0 dB, respectively. These results are a reminder of the Weber law which highlights the proportionality between JND of two stimuli and the magnitude of them. In Figure 6.15 the cumulative loudness distribution is shown for the loudness test with varying SPL.

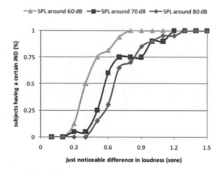

Figure 6.15: Just-noticeable differences in loudness for three sound pressure levels (70 - 80 - 90 dB)

According to this figure, JND in loudness is level-dependent, and with the increase of sound pressure level the recognition of the tiny changes in loudness gets more difficult. The authors have concluded that the loudness changes of 0.8 sone are perceivable for 75% of subjects, who listen to the signal around 70 and 80 dB, and loudness of 0.5 *sone* for the signal around 60 dB.

The deviation of perceived loudness is calculated in percentages with respect to the reference signal with 60, 70 and 80 dB as 2.4%, 4.4% and 4.8 %, thus the loudness JND is approximately in the range of $2-5$ % for level changes of 60 to 80 dB.

Sharpness Concept

Sharpness analysis indicates if the spectrogram of the sampled signal has higher amplitude gravity in the high frequencies. Sharpness is calculated according to (FASTL and ZWICKER, 2007; PULKKI and KARJALAINEN, 2015):

$$S = 0.11 \frac{\int\limits_{0}^{24Bark} N'(z)g(z)z}{\int\limits_{0}^{24Bark} N'(z)dz} \tag{6.9}$$

$N'(z)$ indicates the specific loudness and $g(z)$ is a gain factor on the Bark scale of z and plotted in Figure 6.16: According to the sharpness equation and the gain function, sharpness shows

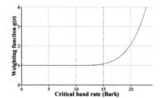

Figure 6.16: Gain function of sharpness

a small dependency on the level and bandwidth (as long as it is smaller than the critical band). Sharpness has a linear relation with the critical band rate under 3 kHz and above that sharpness increases faster which leads to the higher sharpness sensation. In sharpness analysis, the spectral contents and center frequency of the narrow band noise are significant.

Just-Noticeable Differences in Sharpness

A JND in sharpness is investigated by the same author who worked on JND in loudness with the same method and reference stimuli (PEDRIELLI et al., 2008). The authors have changed the sharpness of three reference signals with ±0.18 $acum$ with step size of 0.02 $acum$. The reference signals have original sharpness of 1.42, 1.47 and 1.49 $acum$ for samples with SPL of 59.1, 69.0 and 78.9dB, successively. The result of the listening test for JND in sharpness is shown in Figure 6.17:

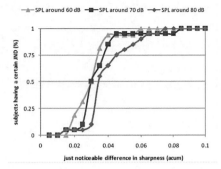

Figure 6.17: Just-noticeable differences in sharpness for three sound pressure levels (70 - 80 - 90 dB) adopted from PEDRIELLI et al., 2008

The authors concluded that the sharpness deviation of 0.04 $acum$ is perceivable; however, it is not level dependent. The perceivable sharpness with respect to reference signals is around 2.6%.

Fluctuation Strength and Roughness Concepts

Fluctuation strength and roughness are phenomena that involve modulation of sound; when the modulation is in the lower frequencies up to 16 Hz, it is called fluctuation strength and above this frequency range is considered as roughness. The unit of fluctuation strength is vacil and it is described as 1 kHz tone at 60 dB with amplitude modulation at 4 Hz. Fluctuation strength shows a peak at modulation frequency of 4 Hz which means higher correlation with the hearing system (easy to be recognized). It also increases with increasing sound pressure level (2.5 vacil with 40 dB level) FASTL and ZWICKER, 2007. Modulation depth ΔL and calibration factor (cal) influence the fluctuation strength as well. Fluctuation strength decreases with an increase in the center frequency for the signals with amplitude modulations.

$$F[vacil] = \frac{0.008 \int\limits_{0}^{24Bark} (\Delta L(z))dz}{\left(\frac{f_{mod}}{4[Hz]}\right) + \left(\frac{4[Hz]}{f_{mod}}\right)} \qquad (6.10)$$

When the modulation frequency of the signal is above 16 Hz, the hearing system doesn't recognize the modulation frequency and roughness occurs. Roughness is calculated as:

$$R[asper] = cal \int\limits_{0}^{24} f_{mod}\Delta L dz \qquad (6.11)$$

Asper is the unit of roughness and is 1-kHz tone which is 100% amplitude modulated at the rate of 70 Hz. The modulation frequency (f_{mod}) and degree of modulation (cal) are important aspects in the roughness calculation. The modulation frequency range in fluctuation strength and roughness is addressed in Figure 6.31.

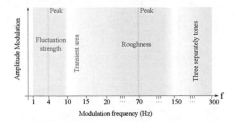

Figure 6.18: Modulation frequency range of fluctuation strength and roughness for a tone signal of 1KHz tone

The just-noticeable differences in fluctuation strength and roughness are given in the next subsection.

Just-Noticeable Differences in Fluctuation Strength and Roughness

A just-noticeable difference in roughness has been investigated by measuring the noise of 12 side-by-side refrigerators (YOU and JEON, 2006). They have changed the fluctuation strength and roughness of a reference signal with amplitude modulation in the time domain. Just-noticeable differences in the fluctuation strength and roughness were measured as 0.01 $vacil$ and 0.03 $asper$. The just-noticeable difference of roughness of 17% was also reported with 10% amplitude modulation (HAVELOCK et al., 2008).

6.2.2 Uncertainty Analysis of Psychoacoustic Parameters

Loudness Uncertainty Analysis

The loudness was calculated according to DIN 45631/A1, 2010, which was explained in subsection 6.2.1, for the uncertainty model. The result of the loudness analysis in time is illustrated in Figure 6.19.

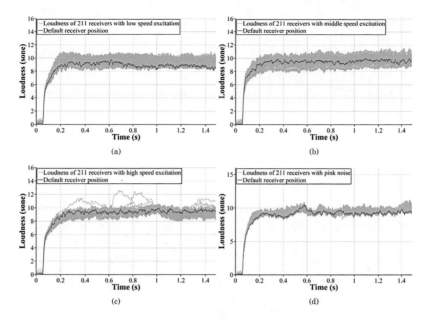

Figure 6.19: Loudness analysis of uncertainty model at (a) low speed, (b) middle speed, (c) high speed and (d) pink noise excitation

In this analysis, the bold plotted line indicates the loudness of the reference sample (Sample No.1) and the cloud of light color indicates the loudness of 211 samples. The loudness analysis of four excitation groups shows deviation around 10 sone and the system with pink noise excitation shows less deviation width in comparison with the engine noise excitations (Figure 6.19(a), (b), (c)). The relative deviation in loudness in all cases is calculated in time using equation 6.12 and is shown in Figure 6.20.

$$\sigma_{RL}(t) = \frac{\sigma_L}{L_m} * 100 \tag{6.12}$$

where σ_{RP}, σ_P and $\tilde{P}s$ stand for the standard deviation and mean value of the psychoacoustic parameters, like loudness, sharpness, roughness and fluctuation strength.

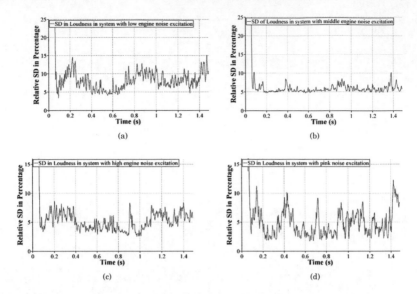

Figure 6.20: Relative SD of loudness analysis in the uncertainty model (in percentages) at (a) low speed, (b) middle speed, (c) high speed and (d) pink noise excitation

The relative SD of loudness shows the minimum loudness deviation of the system with the middle speed noise and the maximum with the pink noise excitation. The loudness of the reference sample, the bold line in Figure 6.20, with each excitation source varies in the cloud of the responses. In order to indicate the noticeable loudness in each sensor position, the 5% percentile loudness (N5) of each sample is calculated. The deviation of N5 is calculated with respect to the N5 of the reference sample and is presented in the vertical axis of Figure 6.21.

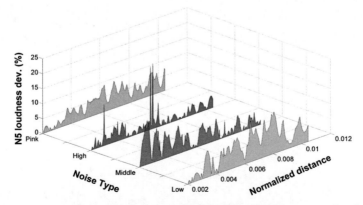

Figure 6.21: Deviation in Loudness N5 in all samples

The distance axis is also normalized to the geometrical length of the system under test. The system with middle speed engine noise shows higher deviation near the reference source which makes the recognition of differences easier; however, for low, high and pink noise samples the deviation is higher in the distance farthest from the reference sample. There are few extreme loud samples in high frequency, however in general, high frequency shows lower deviation in distance. Thus it is expected that the perception of differences with high speed noise will be more difficult compared with middle and low speed engine noise. In the case of pink noise, the deviation in the farthest distance should be easier to recognize in comparison with the samples closest to the reference sample.

Sharpness Uncertainty Analysis

The sharpness analysis is performed on four groups of the uncertainty model and the results are provided in Figure 6.22. Looking at the fundamental frequencies of the engine noise at low, middle and high speed (48 Hz, 179 Hz and 248 Hz), and comparing these with the gain factor of sharpness, Figure 6.16, those frequencies are in the zero slope of the gain factor of sharpness, thus it is expected that the samples are not sharp.

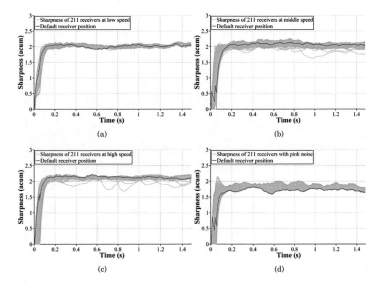

Figure 6.22: Sharpness analysis of uncertainty model at (a) low speed, (b) middle speed, (c) high speed and (d) pink noise excitation

The analysis shows that the sharpness of engine noise is slightly higher than the system with pink noise excitation. The relative deviation in sharpness in each sample is calculated according to Equation 6.12. The result of the relative deviation is shown in percentages in Figure 6.23.

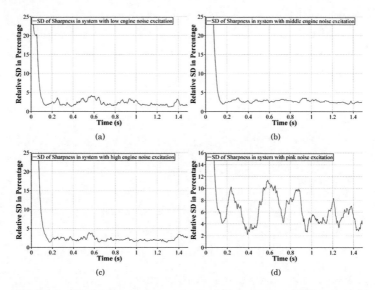

(a)

(b)

(c)

(d)

Figure 6.23: Relative SD of sharpness of uncertainty model in percentages at (a) low speed, (b) middle speed, (c) high speed and (d) pink noise excitation

The samples with engine noise indicate deviation of sharpness around 2.6% in time, so it could be perceivable. In the case of samples with pink noise excitation, the deviations show higher sensitivity in time, despite recognizing the lowest sharpness value with pink noise excitation. The sharpness deviation in each sensor position is obtained via calculation of the single sharpness value and the deviation of each sample from the reference sample. This relative deviation analysis is obtained and shown in Figure 6.24. According to JND in sharpness, sharpness deviation above 2.6% is perceivable. The samples with tonal engine noise show deviation near 2.6%; it can be concluded that the sharpness variation is minor. The samples with pink noise excitation show increases of deviation in distance, thus the samples near the reference one will not be perceivable.

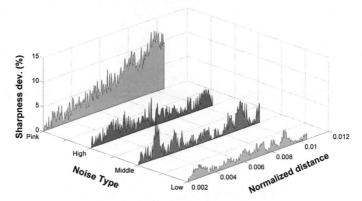

Figure 6.24: Deviation of sharpness in all samples

Fluctuation Strength Uncertainty Analysis

The same pattern of calculating the loudness and sharpness is followed here. First a fluctuation strength is calculated in time in Figure 6.25.

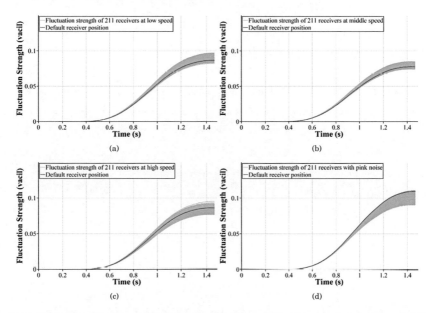

Figure 6.25: Fluctuation Strength analysis of uncertainty model at (a) low speed, (b) middle speed, (c) high speed and (d) pink noise excitation

The Relative SD of the system at all excitations is calculated using Equation 6.12 and plotted in Figure 6.26.

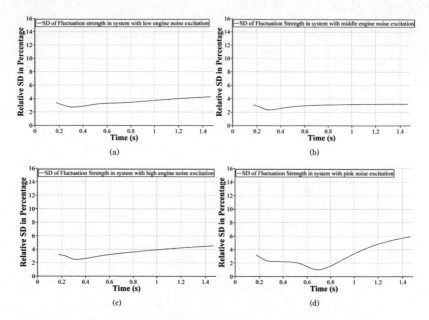

Figure 6.26: Relative SD of Fluctuation Strength of uncertainty model in percentages at (a) low speed, (b) middle speed, (c) high speed and (d) pink noise excitation

A single value of fluctuation strength at each sample position is calculated and plotted in Figure 6.27. This analysis imparts high deviation of the fluctuation strength. In low and middle speed excitations fluctuation strength indicates repeated oscillations in position. In high speed the excitation deviation of fluctuation strength oscillates with larger amplitude, while in pink noise the excitation deviation of fluctuation strength decreases as the position approaching the reference sample.

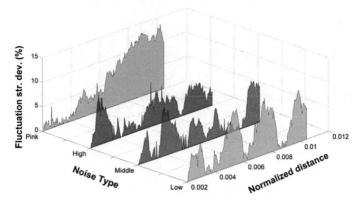

Figure 6.27: Deviation of Fluctuation strength in all samples

The samples with pink noise show a very high deviation of fluctuation strength in comparison with the reference sample. In the literature, there are insufficient reports of noticeable differences of fluctuation strength, thus it is not clear what conclusion can be drawn from this analysis that would allow for it to be linked to the listening test.

Roughness Uncertainties Analysis

The roughness analysis is shown in Figure 6.28 and the relative SD analysis according to equation 6.12 in Figure 6.28:

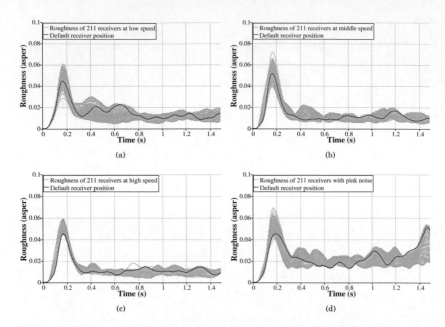

Figure 6.28: Roughness analysis of uncertainty model at low speed (a), middle speed (b), high speed and (c) pink noise excitation (d)

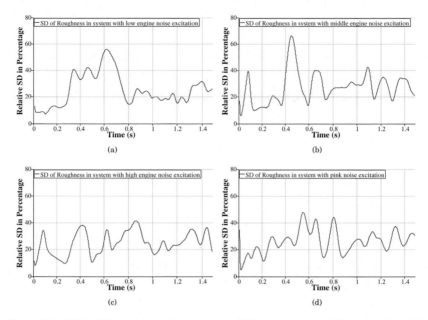

Figure 6.29: Relative SD of roughness of uncertainty model in percentages at (a) low speed, (b) middle speed, (c) high speed and (d) pink noise excitation

The deviation of the single value of the roughness compared with the reference sample is calculated for each distance and illustrated in Figure 6.30.

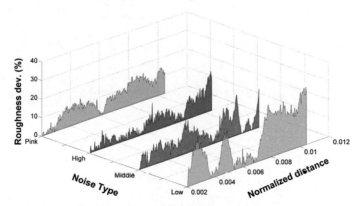

Figure 6.30: Deviation of roughness in all samples

The literature about noticeable roughness is not reported well. Thus, the same conclusion can also be drawn here as for the fluctuation strength.

6.3 Listening Test

A listening test is conducted to specify the perceptual sensitivities of the human auditory system to the sensor positioning uncertainties in measurement of the system with tonal and broadband excitations (Figure 6.1). In this regard, the listening test is developed in two parts: Firstly, threshold analysis is carried out to indicate the minimum perceivable distance differences; secondly, semantic analysis is conducted where the participants describe in which way they perceive the different sound. The two parts are analyzed individually and the results are linked to the sound quality metrics of each stimulus. Before proceeding with the listening test, the research questions should be defined.

6.3.1 Listening Test Question

The alternative hypotheses of the listening test are developed in the following questions:

1. What are the effects of sensor positioning on sound perception?

   Hypothesis[2] : Listeners could perceive the distance deviation.

2. What are the uncertainty perception differences in the system with tonal and broadband excitation?

   Hypothesis: Listeners perceive the best deviation of displacement in a system excited with tonal noise in compare with broadband excitation.

3. Could expert listeners with a musical background perceive uncertainties in sensor positioning better than normal participants?

   Hypothesis: Expert listeners perceive uncertainties more precisely than the normal listeners.

4. How do the listeners distinguish deviation in semantic identification in each stimulus (low, middle, high engine speed noise and pink noise excitation)? Is there any difference between the perception of experts and the normal listener?

   Hypothesis: It is expected that listeners are able to distinguish uncertainties in the system with high speed excitation with loudness deviation, and the system with low speed excitation with roughness deviation. The effects of variation in the system

---

[2]H_1: Alternative hypothesis

excited with middle speed and pink noise are unknown and need to be discovered in the listening test. Also it is expected that experts could explain the deviation better than normal participants.

The first three research questions are investigated with a Bayesian A-B listening test method (two-alternative forced choice), (SREDNICKI, 1988), and the last question is answered by conducting a semantic listening test (ALTINSOY and JEKOSCH, 2012). Both experiments are implemented simultaneously while the semantic section was optional for the listeners to answer.

6.3.2 Listening Test Methodology

The A-B listening test method is developed to indicate the threshold of human hearing to the uncertainties in sensor positioning. In this method, listeners hear three sound events, namely A-B-X, and the X sample is randomized between A and B, so two samples are identical and one different. The participants should pick out the different sample. In this method, the probability of C correct answer with N trials is explained as:

$$P(C|N,p) = \frac{N!}{(N-C)!C!}p^C(1-p) \tag{6.13}$$

where p is the probability of correctly identifying the different sample. This equation is also known as Bernoulli trials (PAPOULIS, 1984) which gives two possible responses of "success" or "failure". Thus the probability of the correct answer is calculated as:

$$p = h + \frac{1}{2}(1-h) = \frac{1}{2}(1+h) \tag{6.14}$$

where h is the fix fraction of N trials. Using this method the probability of having a correct guess is 0.5. Two samples in this test are always the reference sound and the third one is taken from the receiver line in Figure 6.2. The distance between the reference sample position and played sample increases or decreases according to the response of the participants. There are two psychophysical methods to update this distance: classical and adaptive methods. The adaptive method is more plausible since the saturation can be controlled with designed parameters.

In this listening test, a Quest method, the adaptive procedure is adapted to update the threshold analysis. Based on this method, two assumptions are always made (WATSON and PELLI, 1983):

1. "The psychometric function has the same shape under all conditions when expressed as a function of log intensity. From condition to condition, it differs only in position along the

log intensity axis. This position is set by a parameter, T the threshold, also expressed in units of log intensity".

2. "The parameter T of the psychometric function [3] does not vary from trial to trials"

For the Quest method, a probability density function (PDF) is assigned to the physical stimulus as shown in Figure 6.34, and one to each "Yes" and "No" answer of participants. A multiplication of these two PDFs creates a new PDF. The new PDF applies a new threshold to the distance. The threshold distance gets closer or farther to the reference sample with respect to the answer of participants. This iteration continues at least 20 times.

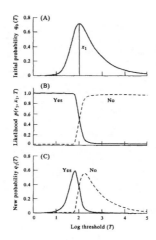

Figure 6.31: Implementation of quest method adapted from KING-SMITH et al., 1994

The functionality of algorithm is evaluated in extreme cases such as when all the answers are correct or false. The waterfall plot of all 20 PDFs in each extreme case is plotted in Figure 6.34:

---

[3] Psychometric function describes the relation between some physical measure of a stimuli and the probability of a particular psycho acoustic response.

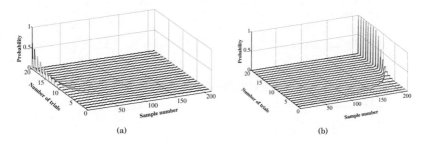

Figure 6.32: PDF of Quest updates (a) all correct answers, (b) all wrong answers

The peak of PDFs in each trial is moving to the left for correct answers, near the reference sample, or to the right with false answers. Readers are referred to the literature for more explanation of the method, (KING-SMITH et al., 1994, WATSON and PELLI, 1983).

In this research, 70 participants were selected with 35 persons with normal hearing abilities and 35 people with expertise like playing instruments or singing in a choir.

Acoustic Stimuli

A total of 80 acoustic stimuli were utilized in this listening test, these samples were chosen from four groups of acoustic conditions: Recorded sound in an uncertainty system excited with electric engine noise at low, middle and high speed and pink noise, respectively. Each group consists of 20 samples taken from a receiver line in Figure 6.2. All 80 samples are randomized to prevent any off-set to the response of participants. A short pause with 30 seconds was scheduled after the first 40 questions and the listeners were questioned to evaluate the test and whether it was difficult or easy to answer. The time duration was recorded in both sections as well.

6.3.3 Listening Test Setup

The listening test was conducted in a hearing booth at the Institute of Technical Acoustics (ITA), RWTH Aachen University. The listening test setup is shown as a block diagram in Figure 6.35, each block is indicated with units.

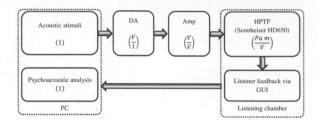

Figure 6.33: Block diagram of the listening test implementation

The acoustic stimuli, which were obtained in the measurement procedure in section 6.1.1, were played back to the participants via audio Interface to the headphone inside the listening chamber. A Graphical User Interface (GUI) was developed to play the stimuli and semantic part of the listening test.

Listening Chamber

The listening chamber has a floor area of 6.8382 m^2 with a ceiling height of 1.98 m. The room is designed with a door and window, Figure 6.34. The background noise level is measured during daytime (workday) with $LAeq_{40min} = 35.4$ dBA measured with sound level meter (Norsonic TYP. 116, KL. 1).

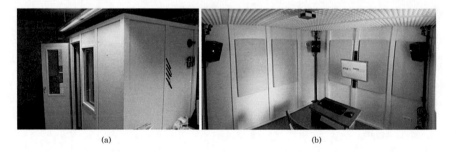

(a)   (b)

Figure 6.34: Auditory booth (a) external view, (b) internal view

Figure 6.35: Background noise measurement inside the auditory booth

Graphical User Interface

The graphical user interface (GUI) was developed to implement the A-B listening test method and the semantic method. The test starts by pushing the start button and participants are guided to listen to the reference sound first, then sound A and B. They were asked to select the sound that seems different from the reference one. In the second part, listeners could activate the semantic part of the test and write in their own words how they perceive the difference between two samples. The next button was designed to continue with the next question.

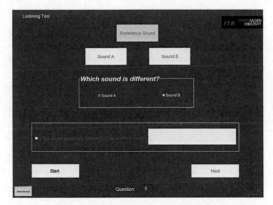

Figure 6.36: Graphical user interface of the listening test

6.3.4 Audiometry

The hearing ability of each participant was examined by using an audiometry test. The test took place at an auditory room of the Institute of Technical Acoustics (ITA) with audiometric Headphone to prevent any exterior interventions. A well accepted method (bracketing) was used to test the audibility of the participants, in which the participant listened to pure tone with different frequencies and sound pressures. The pulsation was generated with increasing frequency within audible range. The level of each pulse increased with time, and the participant was supposed to push a button to indicate if he/she hears the sound. The button should be pushed as long as he/she hears the tone, and then released when the tone is not audible. This push-and-release method determined the hearing threshold of the participant at a certain frequency and this cycle was repeated for certain other frequencies. This examination took approximately 12-13 minutes. All the 70 participants showed normal hearing thresholds. A sample of the Audiometry analysis is shown in Figure 6.37. In this graph, the hearing level in the right and left ear is shown with red and blue colors; the normal hearing threshold at 20 dB is shown with a green broken line. It should also be noted here that, for simplified handling, the measured values of an audiometry are offset with the resting threshold, which is anchored in the international standard, so that a straight line is expected at $0dB$ for normal hearers. The Audiogram thus shows the deviations from the standardized curve for normal hearing.

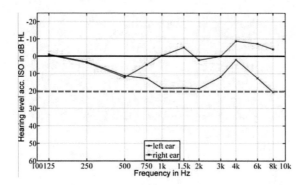

Figure 6.37: A sample of audiometry analysis

Listener Participants

All the 70 participants were recruited by the University of RWTH Aachen at the Institute of Technical Acoustics (ITA) and were classified in two groups of normal and experts; the normal group consists of people with normal hearing abilities without any special auditory training and the experts are participants who play instruments or sing in a choir. The purpose of splitting the participants into these groups is to study the influences upon individuals in their recognition of events referred to in the questions. The listener profile is shown in Table 6.2, referred to in the questions.

Table 6.2: The profile of participants in the listening test

	Men	Women	Age range	Total Number
Normal participants	17	18	19 - 36	35
Expert participants	22	13	18 - 42	35

6.3.5 Evaluation of the Listening Test

The collected data in the listening test are analyzed according to the research questions provided in subsection 6.3.1. In the threshold analysis part, first the threshold result for each stimulus is visualized using boxplots, then the detected thresholds are statistically evaluated using a Repeated Measures Design (ANOVA). A statistical dependency is identified between probability of perception and psychoacoustic parameters of the played stimuli.

In a semantic part of the test, a semantic space of the test is created according to the optional response of the listeners. Additionally, a possible correlation between the semantic space and the psychoacoustic parameters is also discussed.

At the end, the recorded time period of the test is statistically evaluated and connected to the evaluation of the test, and whether it was hard or difficult to recognize the different samples.

Boxplot Analysis

A box-wisker diagram (boxplot) is a method to visualize the data in terms of their quartile by arranging a set of data from the lowest to the highest rate FIELD, 2013. The first quartile is assigned as 25 percentile of data Q_1, the second quartile (median) is 50 percentile $Q2$, and consequently $Q3$ is 75 percentile of the data observation. The outstanding data with large distance from population (outliers) is indicated with dots. A sample of the boxplot with the defined areas is shown in Figure 6.38.

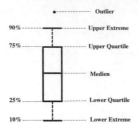

Figure 6.38: 1D Boxplot description

In the listening test each participant answered 80 questions consisting of 20 from each group of stimulus to identify the minimum distance perception in each category; samples with engine noise at low, middle and high speed and pink noise. The boxplot analysis is utilized to illustrate the threshold of distance deviation recognition for a total of 70 participants in Figure 6.39. In this boxplot, the median line of the samples with middle and high speed noise is best recognized in comparison with low speed and pink noise for all participants. The participants show minimum distance deviation in comparison with the four quartiles of the threshold recognition.

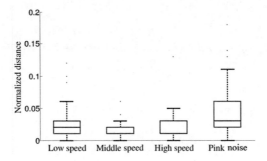

Figure 6.39: Boxplot analysis of all participants

The boxplot analysis is carried out individually for the participants in the normal and expert groups and shown in Figure 6.40.

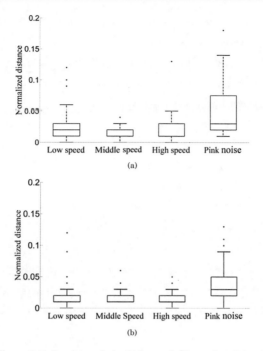

(a)

(b)

Figure 6.40: Boxplot analysis of (a) normal, (b) expert participants

To assess the interaction between the two groups of normal and experts the mean of the threshold in each stimulus category is evaluated and plotted in Figure 6.41.

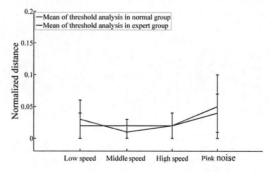

Figure 6.41: The interaction graphs between mean of threshold detection in normal and expert groups

According to the analysis, two groups of participants might have a significant interaction, since two lines in the interaction graph cross each other. The majority of the expert group shows a better distance resolution in each stimulus group in comparison with the normal one; however, in the middle speed group, the normal participant group recognizes the distance deviation better. However, this analysis is not sufficient to claim the statement with certainty, and a more accurate analysis like, repeated measure analysis of variances (ANOVA), is needed to check the possibility of a significant interaction. The plot also reveals that both participant groups indicate the higher median in the distance recognition of the sample with pink noise excitation, which shows the difficulty of both groups in recognition of the broadband excitation in comparison with the tonal excitations (samples with low, middle and high engine speed excitation). Despite the large deviation in recognition of the broadband and tonal excitation, it is interesting to look at the threshold of tonal excitation while considering the fundamental frequency of the tonal noise at each engine speed. To link the recognition of the threshold in each tonal noise group, the fundamental frequency of the engine noise at each speed is obtained and the normalized threshold is calculated based on the normalized wavelength of the engine noise and geometrical length of the system under test. The engine noise at three speeds has the following fundamental frequencies: $48Hz$, $179Hz$ and $248Hz$, respectively, the normalized wavelength is calculated by considering the geometrical length of the system and the fundamental frequency of the excitation signal:

$$\lambda_n = \frac{\lambda_{i1}}{L_g}, \ \lambda_{i1} = \frac{c}{f_{i1}} \tag{6.15}$$

where L_g represents the geometrical length of the system under test and λ_{i1} is the fundamental wavelength of the engine noise at i^{th} engine speed. A boxplot of tonal engine noise corresponding to the fundamental frequency of engine noise excitation is plotted in Figure 6.42.

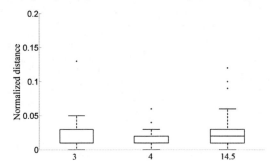

Figure 6.42: The box plot of low, middle and high speed excitation

It can be seen that the threshold of the distance deviation rises with the increase of the fundamental wavelength of the excitation signal with respect to the geometrical length. The results can be interpreted in a way that the uncertainties in a system with tonal excitation at the higher frequency can be recognized better in comparison with the lower frequencies.

In order to look at the data more precisely, repeated measure analysis of variances (ANOVA) has been applied to the data analysis. The method is utilized to indicate the significant threshold differences between each group of participants, and internal deviation of threshold recognition in each stimulus group.

Repeated Measures Design (ANOVA)

The repeated measurement analysis applies to the experiments with the same experimental conditions for each participant at different points of time (FIELD, 2013). In this method, the influence of each independent variable on the dependent variable could be recognized by creating a linear equation involving all variables and possible interaction between them. The following equation is shaped to analyze the results in this work:

$$Threshold = \beta_0 + \beta_1 LT + \beta_2 NT + \beta_3 NT \times LT \qquad (6.16)$$

The independent variables are the listener and noise type (LT, NT), respectively. The interaction between these two variables is indicated with $\beta_3 NT \times LT$. The dependent variable is the threshold of the distance deviation perception; $Threshold$.

The ANOVA analysis indicates that the type of listener with $F(1,67) = 1.45$, gives $p = 0.23$, p is more than 0.05, thus the effect of expert and normal participants is not significant despite the possible interaction report, which was obtained in subsection 6.3.5. The noise type with $F(3,201) = 8.74$, gives $p = 0.17 * 10^{(-12)}$. It means that the noise type has a significant influence on the results. The adjustment with Greenhaus-Geisser gives a new p $p < 0.01(p = 0.3 * 10^{(-2)}), n)^2 = 0.008$. The final group is the interaction between listener type and noise type, again the analysis with $F(3,201) = 1.64, p = 0.18$ shows that this interaction is insignificant ($p > 0.05$). It is crucial to mention that the interaction of listener groups is insignificant, thus further threshold analysis of participants could be continued with the total number of the groups without adding extra weight to the sampled participants. The only significant variable in equation 6.16 is noise type. The noise type consists of four independent variables; samples with engine noise at three speeds and pink noise, and a depended variable is a threshold recognition. A one-way repeated measure ANOVA is adjusted to ascertain the following test groups:

1. Comparison of tonal noise (low, middle and high) versus broad-band noise (pink noise) excitation.

2. Comparison of low speed noise against high speed engine noise excitation.

3. Comparison of middle noise versus high speed engine noise excitation.

These analyses indicate that the influence of all groups is significant with the following p-values: $p_1 = 0.000001, p_2 = 0.000001, p_3 = 0.0000029$, respectively. All the p-values are less than 0.05 which highlights significant differences in each analysis group.

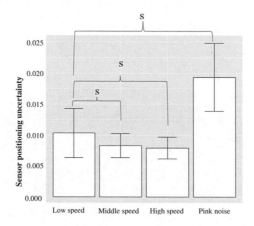

Figure 6.43: Error analysis of the listening test

Referring to the research questions, the first hypothesis is approved, however, it is dependent on the type of the excitation source, thus the second null hypothesis is rejected. The participants could perceive the mean distance deviation up to 2 percent in a system with tonal excitations and near 5 percent with the broadband noise excitation. The third hypothesis is rejected since the type of listener has no significant influence on the perception of the distance deviation. The expert participant could perceive the distance deviation slightly better, however, it is not significant. The fourth hypothesis can be addressed by analyzing the semantic space of the difference perception in the next subsections.

Semantic Test

The verbal description of the perceived differences in each participant group is rated under three psychoacoustic parameters (sound quality metrics): Loudness, sharpness and roughness. A sample of the description is provided in Table 6.3. This table shows the deviation of the given words by participants to describe the difference in each comparison.

Table 6.3: Correlation between answers and psychoacoustics parameters

Loudness	Sharpness	Roughness
Louder/silently	Sharp	Turbulent
Quiet	Damping	Oscillating
Far away/Reverb	Frequency/Freq range	New tone/Harmonic
Less dB	High pitch	Interference/Resonance
	Shrill	Chorus/Out of tune
	Deeper	Vibration

In the listening test, 80 pairs of the samples were compared for each participant, and each participant could express the different sample in their own words in the semantic part of the test, thus the total number of 5,600 expressions were possible for all 70 participants. The participants in total gave 787 comments of which the normal groups contributed 476 comments and the experts contributed 311. Table 6.4 shows the number of comments in each stimulus group:

Table 6.4: Number of semantic samples for each stimuli

	All participants	Normal group	Expert group
Low speed	166	99	67
Middle speed	205	129	76
High speed	255	158	97
Pink noise	161	90	71

The semantic space of the listening test under sound quality metrics: Loudness, sharpness, roughness is visualized using a radar plot in Figure 6.44. The total number of people who rated each group of stimuli is shown in Figure 6.44(a). Then the semantic space for normal and expert groups is divided and presented in Figure 6.44(b) and 6.44(c), respectively. In this plot, the number of comments in each sound quality metrics was presented in the radii of the radar plot, and the people who defined the differences as unknown as the four radii of the plot. Four groups of lines can be recognized in each radar plot, which indicates the samples with low, middle and high engine speed noise and pink noise. The magnitude in each group is connected via a line that makes the comparison easier.

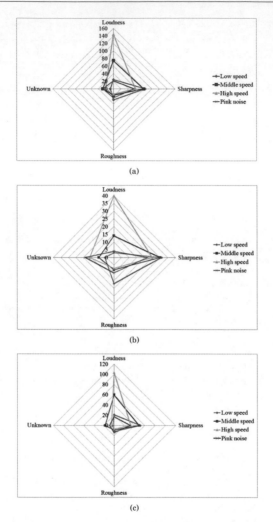

Figure 6.44: Radar plot of semantic space (a) all participants, (b) normal group, (c) expert group

According to Figure 6.44(a), all participants reported samples with low speed excitation with mainly sharpness deviation, and the middle speed excitation with loudness and sharpness. In samples with high speed engine noise, listeners perceive the deviation in sensor positioning with loudness. A few comments are given for pink noise excitation which indicates the difficulties in perception of the sensor positioning with the broadband excitation.

Time and Difficulty

There are 80 questions for each participant to answer, and after the first 40 questions, a pop up has been activated to evaluate the difficulties of the test. The participants had two options to evaluate the test: Easy or difficult. The time interval between the two sections with 40 questions was recorded. The interaction between the time in each section and their evaluation is performed in this section.

Before starting the evaluation, a Grubbs' Test is performed to detect the possible outliers with the following hypothesis

Null hypothesis: All data values come from the same normal population

Alternative hypothesis: Smallest or largest data value is outliers

The results of the test are indicated in Table 6.5:

Table 6.5: Outliers analysis

variable	N	G	p
Easy-1st part	40	2.99	0.061
Easy-2nd part	40	2.71	0.183
Difficult-1st part	8	1.72	0.438
Difficult-2nd part	8	1.55	0.773

According to the p-value there are no outliers at the 5% level of significance. Thus the null hypothesis is valid.

The test duration in the first and second part is compared via the paired t-test. The null hypothesis says that the mean of time duration for all participants is similar in the first and second part of the test. According to the analysis, $p-value$ is 0.03, which is at the default 5% significance level, and rejects the null hypothesis. Thus the mean of the samples is different. This is a meaningful deviation, and highlights the difficulties of distance recognition near the default receiver, despite the accurate deviation recognition according to the threshold analysis. The confidence interval is also reported between 0.2051 and 5.2116 min.

Table 6.6: Listening test duration analysis

variable	Mean (min)	25 percentile (min)	50 percentile (min)	75 percentile (min)
1st part	12.3	8	10	16
2nd part	15.1	9	12.5	19

In total, 83% of participants reported that the first part was easy to recognize and 17% reported this as difficult, and all the participants reported that the second part of the test was difficult in comparison with the first part. Finally, it can be concluded that distance recognition in 21% of the normalized distance is easy to recognize and, above this threshold, recognition is possible, but it takes a longer time to be recognized.

6.4 Summary

The influence of the sensor positioning on the sound design parameters was investigated in this chapter. Perception of the sound field was evaluated with objective and subjective considerations; the objective evaluation was considered by calculation of the sound quality metrics and subjective evaluation by conducting a listening test. The relative deviation of the sound quality metrics (loudness, sharpness, fluctuation strength and roughness) was obtained. It was stated that loudness deviation was the most influenced parameter, according to the JND in loudness. The sharpness deviation was minor except for the pink noise sample.

The listening test was conducted in two sections of threshold for distance uncertainty recognitions and the semantic part. The threshold analysis was performed to indicate the influence of the type of input excitation on the recognition of the distance deviation on two groups of normal and expert participants. The interaction between each group was studied as well. The analysis indicated that the type of participants has no significant influence on the threshold recognition of the samples, despite the interaction between the median threshold of the distance recognition between participants. The threshold analysis in the sensor displacement analysis indicated that samples with broadband excitation are more difficult to recognize in comparison with samples with tonal excitation. The analysis continued with a close look at the threshold recognition within the samples with tonal excitation. The analysis indicated that the sample with higher fundamental frequency is easier to recognize in comparison with the lower fundamental frequency.

In the last step, the first and second time intervals created in the test were compared. The result was given that the participants could recognize the first half of the samples faster than the second part, and it shows that they needed more concentration and time to recognize the deviation. Thus, the real distance recognition is time dependent, and this should be taken into consideration for the application of the sound design medium.

The method proposed in this chapter, based on the objective and subjective uncertainty evaluation of the sound field, could be applied to the practical case studies.

7

Application in Automotive Acoustics

In this chapter, the application of the uncertainty analysis is presented in vehicle acoustics. An electric vehicle is being tested and a transfer path is defined between an excitation source, an engine mount and a receiver – a passenger compartment. The uncertainties are defined by measuring the position inside the vehicle cabin. The measurement setup is explained below, and then the uncertainty is introduced to the system. Ultimately, the uncertainty analysis method is applied to the measured results.

The measurement of the electric vehicle is performed in a semi-anechoic chamber at the Institute of Technical Acoustic (ITA), RWTH Aachen University. The electric vehicle being tested has dimensions of $x = 2.69\ m$, $y = 1.38\ m$ and $z = 1.54\ m$ respectively, Figure 7.1. A shaker has been placed under the drive train support frame, Figure 7.2(b), to excite the system, and a microphone is connected to the steering wheel of the electric vehicle via a mechanical setup, Figure 7.2(a), to measure the sound field inside the passenger compartment. The mechanical setup is designed to keep the microphone in the desired measurement positions with capability of movement in x and y direction, the horizontal plane. Moreover, the mechanical setup is designed to rotate around the joint of the microphone connection in the vertical plane.

A reference measurement point is defined at 660 mm and 920 mm in x and y direction at the furthest point from the extreme right-hand side of the passenger compartment. The microphone is displaced from a reference point at x and y direction with the step size of 1 mm up to 81 mm, set 1 and set 2, Figure 7.2(c). Furthermore, the sound field is measured in the diagonal displacement, which starts from the last measured position in the x axis, and ends at the last measured position in y direction; in this measurement the first record is defined as the reference point. The step size in this set is 17.5 mm and the total displacement of 114.6 mm is obtained, set 3, Figure 7.2(c). Considering the original position of the microphone at 90° in the vertical plane, Figure 7.2(d), the measurement is made at 120°, 90°, 80°, 60°, 50°, 40°, 30°, −2°, −10°, −20°, 30° and −75° positions, respectively, this set of measurements is named as set 4. In this set, the maximum sensor displacement from the reference microphone position is 220 mm. The width of uncertainty in sensor positioning could be reported with two sets of 81 mm, a set with 1146 mm and the last set with 220 mm.

Figure 7.1: Electric vehicle in the semi-anechoic chamber

A sweep signal from 10 Hz to 20 kHz is generated to excite the system with the shaker ($B\&K$, type 4809), and the microphone (Sennheiser, KE4) is utilized to measure the sound field inside the passenger cabin. The measurement chain is able to produce a linear behavior up to 5 kHz, the restriction is due to limitation of the shaker to produce a linear force above 5 kHz. The measurement setup remained the same during the sound field measurement for all sets. The measured transfer paths in each set are shown in Figure 7.3. In this figure, the bold line indicates the reference transfer path and the light plots are the number of the measured positions in each set.

The uncertainties in each set of the measurement are evaluated using the proposed uncertainty method in chapter 3. Hence, a relative SD of each set is calculated, and the averaged power of them is obtained and plotted in Figure 7.4. The x axis in this plot is the normalized frequency based on the geometrical length of the vehicle and the measured frequency range.

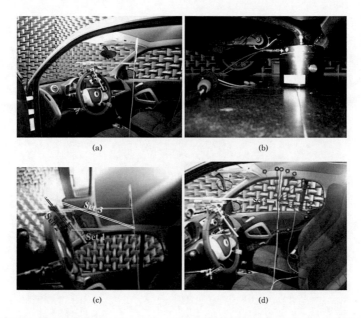

(a) (b)

(c) (d)

Figure 7.2: Measurement setup: (a) the position of microphone via a mechanical setup inside the vehicle (b) a shaker to excite the system connected to the drive train support frame (c) measurement set 1, 2 and 3, (d) measurement set 4

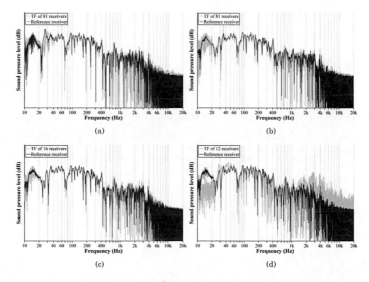

(a) (b)

(c) (d)

Figure 7.3: Transfer path measurement with sensor positioning in (a) set 1(b) set 2 (c) set 3 (d) set 4

Considering the uncertainty results in Figure 7.4, set 1, 2 and 3 show the similar deviation range in the frequency range of interest; while set 4 shows a higher uncertainty power. This is due to the large displacement from the reference sample. The trend of error increases in the frequency domain is the same as it was for the simple system described in Chapter 3.

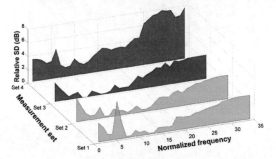

Figure 7.4: Smart vehicle

This experiment supports the theory proposed in chapter 3 stating that with the sensor displacement of up to 6% of the geometrical length of the system from the original position, the deviation of the measured transfer path remains below 4 dB. In this case, the geometrical length of the system is about 1.8 m and the relative displacement in the first and second sets are 4.5% and the third set is 6.3%. The last set has a relative displacement of 12.3% which is twice as large as the proposed sensor positioning uncertainty. Referring to the results in Figure 3, the uncertainties around 6% show the maximum of 4 dB, and set 4 shows exactly twice the level of uncertainties.

This result can be used as a guideline for measurement of a transfer path in complex systems. Moreover, this investigation facilitates research into modeling of vibro-acoustic systems, which is very complicated in the higher frequencies, thus the simplified models could be utilized in the general estimation of the transfer path by considering the maximum deviation.

8

Conclusion and Outlook

This study presented the discussion of uncertainty modeling and analysis of a transfer path. Therefore, this research introduced samples of acoustic and structural transfer paths associated with internal and external sources of uncertainties. Then, the systems were analytically modeled and a relative standard deviation analysis was applied to the uncertainty models.

The result of the uncertainty analysis in the acoustic transfer path was shown as a trend of deviation increase in the frequency range of interest and a saturation above the Schroeder frequency. The relative SD was shown as a saturation at 4 dB for the acoustic system with the external uncertainty sources, and 6 dB with the internal source. Furthermore, the influence of the tonal excitation on the error propagation was calculated.

In the structural transfer path, first the overlap frequency in a sampled structure was calculated, then the uncertainty analysis was applied to the structural transfer path with the sensor positioning uncertainties. The calculated relative standard deviation of error was shown as a high deviation in the lower frequencies and a slight increase of error up to the overlap frequency, above the overlap region, a flat error response was obtained. Thus, it can be concluded that the structural analysis is sensitive to the sensor positioning uncertainties in the entire frequency range, which is about 4 dB. The uncertainty analysis with a tonal input excitation in the structure shows a similar behavior error in comparison with the broadband input excitation.

The error saturation at the diffuse field was examined using a statistical approach, a K–S test. This approach revealed the significance of the deviation in two PDFs. The K–S test was applied to all the uncertainty analysis cases, the K–S test showed that all the PDFs were significantly different at 95%, except the temperature analysis. In the acoustic system, the temperature changes below $10°C$ showed insignificant differences in the entire frequency range.

The acoustic and structural transfer paths with uncertainty sources were experimentally implemented according to the analytical models. The relative SD of the transfer path with uncertainties was obtained, and the results were also shown for saturation of error. Thus, the statistical assessment was carried out for the experimental results in the overlap region. The analyses indicated significant differences in the Rayleigh distributions for all the cases; however, the Rayleigh fit shows a narrow deviation of the means for the system with uncertainties in the sensor positioning and temperature variation. The system with damping uncertainties showed a large deviation of mean and variances of the Rayleigh fit in the frequencies above the overlap region. Moreover, the acoustic and structural transfer paths in the vibro-acoustic system were measured with the sensor positioning uncertainties. The error analysis indicated that the error saturation shifted back in the frequency according to the type of measured transfer paths.

The averaged power of relative SD in the analytical and experimental analysis was compared to evaluate the analytical uncertainty model. A very good agreement was achieved with the comparison of error analysis. Due to the validity of the analytical model, the uncertainty model was generalized according to the geometrical length of the system being tested.

The uncertainty analysis is applied to a sound design task. In this respect, deviation of psychoacoustic parameters was calculated based on the four types of input excitations. The analysis results were compared with the just-noticeable differences (JND) in each case. Loudness is the most relevant parameter in recognition of deviation in most cases. The deviation analysis is given in time and space to cover all aspects of the deviation.

A study on the perception of deviation recognition with the four types of input excitation source was performed by designing a listening test with 70 participants. The participants were divided into two groups of normal participants and experts, and each participant evaluated the uncertainty deviation with sensor positioning as a source of uncertainty with four types of input excitation; three tonal excitations and a pink noise as a broadband excitation. The results reveal that the expert participants recognize the uncertainties slightly better, however, the deviation in comparison with the normal groups is not significant. As for the recognition of the broadband and tonal excitation, the tonal excitation can be recognized easier by both groups. The internal evaluation of three tonal noises indicates that the tonal noise at higher frequencies has a greater probability of being recognized.

The final attempt in the thesis was to study the application of the proposed uncertainty analysis in vehicle acoustics. It was claimed that the deviation of the uncertainties in the measurement with a maximum of 6% with respect to the geometry length of the system gives a maximum error of 4 dB.

The method of the uncertainty analysis proposed in this dissertation can be applied in the electric vehicle for the purpose of condition monitoring with an assessment of the range of variations due to known or unknown sources of uncertainties.

In the field of sound design, it is recommended to calculate the coloration of sound besides the psychoacoustic parameters, and to apply the relative approach to calculating the pattern changes of the samples.

List of References

ALTINSOY, M. E. and JEKOSCH, U. (2012). "The semantic space of vehicle sounds: developing a semantic differential with regard to customer perception". In: *Journal of the Audio Engineering Society* 60.1/2, pp. 13–20 (cit. on p. 97).

BERANEK, L. L. (1996). "Acoustics". In: (cit. on pp. 7, 13, 127).

BIERMANN, J.-W. (2012). *Vehicle acoustics*. Vol. 16. IKA (cit. on p. 13).

CREMER, L. and HECKL, M. (2013). *Structure-borne sound: structural vibrations and sound radiation at audio frequencies*. Springer Science & Business Media (cit. on pp. 7, 23, 25).

DE VIS, D and HENDRICX, W (1992). "Development and integration of an advanced unified approach to structure borne noise analysis". In: *INTER-NOISE and NOISE-CON Congress and Conference Proceedings*. Vol. 1992. 2. Institute of Noise Control Engineering, pp. 561–564 (cit. on p. 1).

DIETRICH, P; MASIERO, B; LIEVENS, M; VORLÄNDER, M; BISTAFA, S, and GERGES, S. (2010). "Open transfer path measurement round-robin using a simplified measurement object". In: *ISMA Conference on Advanced Acoustics and Vibration Engineering. Leuven, Belgium:[sn]* (cit. on p. 20).

DIETRICH, P. (2013). *Uncertainties in acoustical transfer functions: modeling, measurement and derivation of parameters for airborne and structure-borne sound*. Vol. 16. Logos Verlag Berlin GmbH (cit. on pp. 2, 13, 47, 48).

DIN 45631/A1, D. (2010). "Berechnung des Lautstärkepegels und der Lautheit aus dem Geräuschspektrum". In: *DIN, Berlin (Germany)*, p. 27 (cit. on p. 86).

FAHY, F.; GARDONIO, P., and HAMBRIC, S. (2007). "Sound and structural vibration". In: *Acoustical Society of America Journal* 122, p. 689 (cit. on pp. 7, 10, 11, 23, 131).

FASTL, H. and ZWICKER, E. (2007). *Psychoacoustics. facts and models*. Vol. 22. Springer Science & Business Media (cit. on pp. 80, 83–85).

FIELD, A. (2013). *Discovering statistics using ibm spss statistics*. Sage (cit. on pp. 103, 107).

GAJDÁTSY, P. Á. (2011). *Advanced transfer path analysis methods* (cit. on p. 18).

GUM (2008). "Guide to the Expression of Uncertainty in Measurement, (1995), with Supplement 1, Evaluation of measurement data, JCGM 101: 2008". In: *Organization for Standardization, Geneva, Switzerland* (cit. on pp. 2, 14, 26, 39).

HAVELOCK, D.; KUWANO, S., and VORLÄNDER, M. (2008). *Handbook of signal processing in acoustics*. Springer Science & Business Media (cit. on p. 86).

HECKL, M. and MÜLLER, H. A. (1994). *Taschenbuch der technischen akustik*. Springer-Verlag (cit. on p. 19).

ISO (2008). "98-3/Suppl. 1: 2008 (E) Uncertainty of measurement–Part 3: Guide to the expression of uncertainty in measurement (GUM: 1995) Supplement 1: Propagation of distributions using a Monte Carlo method". In: *ISO, Geneva (Switzerland)*, p. 98 (cit. on p. 15).

ISO 3382-1, I. (2009). "3382-1". In: *Acoustics-Measurement of room acoustic parameters, Performance spaces* (cit. on p. 56).

KING-SMITH, P. E.; GRIGSBY, S. S.; VINGRYS, A. J.; BENES, S. C., and SUPOWIT, A. (1994). "Efficient and unbiased modifications of the QUEST threshold method: theory, simulations, experimental evaluation and practical implementation". In: *Vision research* 34.7, pp. 885–912 (cit. on pp. 98, 99).

KUTTRUFF, H. (2007). *Acoustics: an introduction*. CRC Press (cit. on pp. 7, 13).

KUTTRUFF, H. (2009). *Room acoustics*. CRC Press (cit. on pp. 7, 20–22).

LENNSTRÖM, D. (2013). *Methods for motor noise evaluation and control in electric vehicles*. Luleå tekniska universitet (cit. on p. 3).

MOHAMADY, S.; MÜLLER-TRAPET, M., and VORLÄNDER, M. (2013). "Transfer Path Analysis of Multi-Structure Acoustic Systems Using a Simplified Measurement Object". In: *AIA-DAGA 2013*. 39th annual congress of DEGA and the 40th annual congress of AIA. Merano, Italy (cit. on p. 49).

MÜLLER, S. and MASSARANI, P. (2001). "Transfer-function measurement with sweeps". In: *Journal of the Audio Engineering Society* 49.6, pp. 443–471 (cit. on p. 49).

PAPOULIS, A (1984). "Bernoulli trials". In: *Probability, Random Variables, and Stochastic Processes*. McGraw-Hill New York, pp. 57–63 (cit. on p. 97).

PEDERSEN, T. H. (2008). "The Semantic Space of Sounds". In: *Delta* (cit. on p. 80).

PEDRIELLI, F.; CARLETTI, E., and CASAZZA, C. (2008). "Just noticeable differences of loudness and sharpness for earth moving machines". In: *Journal of the Acoustical Society of America* 123.5, pp. 3164–3164 (cit. on pp. 83–85).

PULKKI, V. and KARJALAINEN, M. (2015). *Communication acoustics: an introduction to speech, audio and psychoacoustics*. John Wiley & Sons (cit. on pp. 80, 81, 84).

RISSLER, K. (2011). "Applying the reciprocal Transfer Path Analysis (TPA) for the airborne sound of power train components". In: (cit. on p. 28).

ROSSING, T. D. *Springer handbook of acoustics*. Springer (cit. on p. 20).

SCHROEDER, M. R. (1996). "The "Schroeder frequency"revisited". In: *The Journal of the Acoustical Society of America* 99.5, pp. 3240–3241 (cit. on p. 33).

SCHROEDER, M. R. (1962). "Frequency-Correlation Functions of Frequency Responses in Rooms". In: *The Journal of the Acoustical Society of America* 34.12, pp. 1819–1823 (cit. on p. 22).

SMITH, C. L. (1999). "Uncertainty propagation using Taylor series expansion and a spreadsheet". In: *Idaho National Engineering Laboratory* (cit. on p. 1).

SREDNICKI, M. (1988). "A Bayesian analysis of AB listening tests". In: *Journal of the Audio Engineering Society* 36.3, pp. 143–146 (cit. on p. 97).

Van der GIET, M. (2011). *Analysis of electromagnetic acoustic noise excitations: a contribution to low-noise design and to the auralization of electrical machines*. Shaker (cit. on p. 18).

VOGT, T.-S. (2006). *Prognose des emittierten luftschalls von motoren mit hilfe zweier methoden zur schallquellenidentifikation*. Shaker (cit. on p. 29).

WATSON, A. B. and PELLI, D. G. (1983). "QUEST: A Bayesian adaptive psychometric method". In: *Perception & psychophysics* 33.2, pp. 113–120 (cit. on pp. 97, 99).

YOU, J. and JEON, J. Y. (2006). "Just noticeable difference of sound quality metrics of refrigerator noise". In: *INTER-NOISE and NOISE-CON Congress and Conference Proceedings*. Vol. 2006. 1. Institute of Noise Control Engineering, pp. 5581–5586 (cit. on p. 86).

A

Appendix

A.1 Deriving Equations

A.1.1 Wave Equation

First, it is necessary to understand the equation of motion BERANEK, 1996: Knowing that $grad\, p = i\frac{\partial p}{\partial x} + j\frac{\partial p}{\partial y} + k\frac{\partial p}{\partial z}$ and with the assumption of small particle velocity q, the rate of changes of particle velocity in time is equal to the gradient of pressure in space:

$$-grad\, p = \rho_0 \frac{\partial q}{\partial t} \tag{A.1}$$

The second essential equation is gas law which says that: Total pressure P in volume V depends on the temperature (T) and mass of gas:

$$PV = RT \tag{A.2}$$

where R is a gas constant whose magnitude depends upon the mass of gas. It should be mentioned that with diffraction and rarefaction of fluid, the wave temperature increases and decreases with the same frequency and phase of pressure; however, the speed of heat exchange between maximum and minimum temperature is so low that in audible frequency is negligible and we could assume constant deviation, thus equation A.2 becomes:

$$PV\gamma = constant \tag{A.3}$$

where γ is the ratio of the specific heat of fluid at constant pressure to the specific heat at constant volume of the fluid. This equation can be expressed as below:

$$\frac{P}{P} - \frac{-\gamma dV}{V} \tag{A.4}$$

with assumption of small changes of incremental pressure p in comparison with undisturbed pressure P_0 and incremental volume τ with undisturbed volume V_0 we have:

$$\frac{1}{P_0}\frac{\partial p}{\partial t} = \frac{\gamma}{V_0}\frac{\partial \tau}{\partial t} \tag{A.5}$$

In the continuity equation:

$$\frac{\partial \tau}{\partial t} = V_0 div q \tag{A.6}$$

where q is the instantaneous particle velocity. This equation means that the mass of gas remains constant and moves according to the displacement vector.

Now three essential equations are in hand to derive the wave equation. The equation A.2 can be written:

$$\frac{\partial p}{\partial t} = \frac{-P_0\gamma}{V_0}\frac{\partial \tau}{\partial t} \tag{A.7}$$

combining with equation A.6 gives:

$$\frac{\partial p}{\partial t} = -P_0\gamma div q \tag{A.8}$$

with differentiation in time:

$$\frac{\partial^2 p}{\partial t^2} = -P_0\gamma div\frac{\partial q}{\partial t} \tag{A.9}$$

divergence of side of equation A.1:

$$-div\big(grad\ p\big) = \rho_0 div\Big(\frac{\partial q}{\partial t}\Big) \implies -\nabla^2 p = \rho_0 div\Big(\frac{\partial q}{\partial t}\Big) \tag{A.10}$$

combining A.9 and A.10 gives wave equation:

$$\nabla^2 p = \frac{\rho_0}{\gamma P_0}\frac{\partial^2 p}{\partial t^2} \tag{A.11}$$

speed of sound is defined as below:

$$c^2 = \frac{\gamma P_0}{\rho_0} \tag{A.12}$$

thus wave equation can be written:

$$\nabla^2 p = \frac{1}{c^2}\frac{\partial^2 p}{\partial t^2} \tag{A.13}$$

A.1.2 Orthogonal Condition and Modal Coefficient

Function $f(x)$ and $g(x)$ are orthogonal if:

$$\int_a^b f(x)g(x) = 0 \tag{A.14}$$

function $f(x)$ and $g(x)$ are mutually orthogonal if:

$$\int_a^b f_i(x)f_j(x) = \begin{cases} 0 & \text{for } i \neq j \\ K & \text{for } i = j \end{cases} \tag{A.15}$$

Process of calculation of modal coefficient:

$$\int_{L_x} X_i^2 \int_{L_y} Y_i^2 \int_{L_z} Z_i^2 = L_x L_y L_z \tag{A.16}$$

Thus the eigenfrequencies are derived as below:

$$\int_{L_x} X_i^2 = \int_{L_x} A_i^2 cos^2\left(\frac{n_{ix}\pi}{L_x}x\right)dx \equiv \frac{A_i^2}{2}\left[L_x + \frac{1}{2}sin(2n_{ix}\pi)\right]\frac{A_{ix}L_x}{2} = L_x \Rightarrow A_i^2 = 2 \tag{A.17}$$

this equation holds for y and z dimensions that leads to the:

$$A_i B_i C_i = \sqrt{8} \tag{A.18}$$

Thus, equation 3.7 can be rewritten as:

$$\psi_i(x,y,z) = \sqrt{8}cos\left(\frac{n_{xi}\pi}{L_x}x\right)cos\left(\frac{n_{yi}\pi}{L_y}y\right)cos\left(\frac{n_{zi}\pi}{L_z}z\right) \tag{A.19}$$

The transfer path of the system by considering the wave synthesis approach is obtained as:

$$\psi_i(x,y,z) = \sqrt{8}cos\left(\frac{n_{xi}\pi}{L_x}x\right)cos\left(\frac{n_{yi}\pi}{L_y}y\right)cos\left(\frac{n_{zi}\pi}{L_z}z\right) \tag{A.20}$$

wave synthesis at sound source and receiver yields derivation of the transfer path in the enclosure as:

$$TF(\omega) = -\frac{4\pi c^2}{V} \sum_i \frac{\psi_i(r)\psi_i(r_0)}{(\omega 2 - \omega_i^2 - 2j\delta_i\omega_i)} \tag{A.21}$$

the damping constant is denoted with δ_i and can be calculated by knowing the reverberation time RT_i of the scenario:

$$\delta_i = \frac{3ln(10)}{RT_i} \tag{A.22}$$

A.2 Mechanical Basics

Some mechanical expressions need to be defined:

Stress: is defined as force per unit area:

$$\sigma = \frac{F}{A} \tag{A.23}$$

and if the stress is not uniform in the surface area we define it as.

$$\sigma = \lim_{\Delta A \to 0} \left(\frac{\Delta F}{\Delta A}\right) \tag{A.24}$$

Normal stress: Tress perpendicular to the cross-section.

Tensile : Member is in tension $\sigma > 0$.

Compressive: Member is in compression $\sigma < 0$.

Homogeneous: Material is the same through the bar.

Prismatic: Cross-section dosn't change along axis of bar.

Poisson ratio (v): Ratio of magnetic of the lateral strain to the longitudinal strain.

$$v = -\frac{\epsilon_{xx}}{\epsilon_{yy}} \tag{A.25}$$

Young's modulus (E): Ratio of longitudinal stress σ_{xx} to the longitudinal strain ϵ_{xx} in a thin uniform bar.

B:

$$B = \frac{E(1-v)}{(1+v)(1-2v)} \tag{A.26}$$

Shear modulus: Ratio of shear stress τ to the shear strain γ and is related to Young's modulus:

$$G = \frac{E}{2(1+v)} \tag{A.27}$$

Elasticity theory in thin plate:

$$\sigma_{xx} = \left[\frac{E}{1-v^2}\right]\frac{\partial \xi}{\partial x} \tag{A.28}$$

this shows the relation between longitudinal stress and longitudinal strain FAHY et al., 2007.

A.3 Listening Test Documents

Studienleitung:
Prof. Dr. rer. nat. Michael Vorländer
Institut für Technische Akustik
RWTH Aachen
Kopernikusstraße 5, 52074 Aachen
Telefon: 0241-80-97985
Email: post@akustik.rwth-aachen.de

Probandeninformation
„Wahrnehmung von Geräuschen elektrischer Maschinen"

Sehr geehrte Damen und Herren,

am Institut für Technische Akustik der RWTH Aachen wird ein Hörversuch zum Thema *„Wahrnehmung von Geräuschen elektrischer Maschinen"* durchgeführt.

Um weitere Informationen zu erhalten, bitten wir Sie, die folgende Studienbeschreibung aufmerksam durchzulesen.

Studienbeschreibung

Im folgenden Experiment *„Wahrnehmung von Geräuschen elektrischer Maschinen"* werden die Unterschiede in der Wahrnehmung zwischen Elektromotor-Geräuschen und Rosa Rauschen verglichen.

Durchführung des Hörversuchs

Der Hörversuch wird in der Hörkabine des Instituts für Technische Akustik, RWTH Aachen University, Kopernikusstraße 5, 52074 Aachen durchgeführt.

Zu Beginn des Versuchs wird ein Hörscreening (Audiometrie) durchgeführt. Anschließend wird der oben beschriebene Versuch zur *„Wahrnehmung von den Geräuschen elektrischer Maschinen"* durchgeführt. Das Experiment dauert insgesamt etwa eine Stunde.

Risiken und Nutzen

Die Ergebnisse sind ausschließlich für Forschungszwecke bestimmt, es besteht kein zusätzlicher Nutzen oder Risiko durch die Teilnahme.

Die Hörkabine ist etwa fünf Quadratmeter groß. Sie hat ein Fenster und eine Tür, die sich von innen öffnen lässt. Falls Sie sich in der Hörkabine unwohl fühlen, können Sie den Versuch jederzeit abbrechen und eigenständig die Kabine verlassen.

Der Pegel der wiedergegebenen Signale überschreitet dabei niemals 65 dB SPL. Jedes Signal hat eine Länge von 1.5 Sekunden. Es kommt somit auch nicht zu einer Dauerbelastung.

Freiwilligkeit und Anonymität

Die Teilnahme an der Studie ist freiwillig. Sie können jederzeit und ohne Angabe von Gründen die Teilnahme an dieser Studie beenden, ohne dass Ihnen daraus Nachteile entstehen. Auch wenn Sie die

Studie vorzeitig abbrechen, haben Sie Anspruch auf eine entsprechende Vergütung für den bis dahin erbrachten Zeitaufwand.

Die gemessenen/aufgenommenen Daten werden nur für wissenschaftliche Zwecke verwendet und werden keinesfalls an Dritte weitergegeben. Eine Identifikation von Personen mithilfe des aufgenommen Tonmaterials ist zu keiner Zeit im Projekt möglich. Alle Daten liegen vollständig anonymisiert vor und werden streng vertraulich behandelt. Des Weiteren wird die Veröffentlichung der Ergebnisse der Studie in anonymisierter Form erfolgen, d.h. ohne dass Ihre Daten Ihrer Person zugeordnet werden können.

Datenschutz

Die Erhebung und Verarbeitung Ihrer Leistungs- und personenbezogenen Daten erfolgt vollständig anonymisiert an der Professur für Technische Akustik der RWTH Aachen unter Verwendung einer Nummer und ohne Angabe Ihres Namens. Ihre Daten sind dann anonymisiert. Damit ist es niemandem mehr möglich, die erhobenen Daten mit Ihrem Namen in Verbindung zu bringen. Die anonymisierten Daten werden mindestens 10 Jahre gespeichert.

Auffällige Befunde

Die audiometrische Untersuchung dient ausschließlich Forschungszwecken. Eine medizinische oder psychologische Beurteilung Ihrer Daten erfolgt nicht. Es könnte uns jedoch ein ungewöhnliches Untersuchungsergebnis im Hörscreening (Audiogramm) auffallen. In diesem Fall werden wir Sie darüber informieren und Ihnen empfehlen, dieses Ergebnis bei Ihrem Hausarzt oder einem Hals-Nasen-Ohren (HNO) Arzt diagnostisch weiter abklären zu lassen. Nur wenn Sie damit einverstanden sind, dass wir Sie ggf. über einen auffälligen Befund informieren, können Sie an dieser Studie teilnehmen. Sofern bei dieser diagnostischen Abklärung eine Erkrankung festgestellt werden sollte, könnten Ihnen daraus unter Umständen Nachteile entstehen, z. B. der Abschluss einer privaten Krankenversicherung oder einer Lebensversicherung erschwert werden.

Vergütung

Für die Teilnahme an der Untersuchung erhalten Sie eine Vergütung in Höhe von 8 € pro Stunde. Die Vergütung wird Ihnen in bar ausgezahlt. Bei Empfang der Vergütung in bar müssen Sie eine Quittung mit Angabe Ihres Namens und Ihrer Adresse unterschreiben. Diese Informationen werden völlig separat von den Untersuchungsdaten aufbewahrt.

Wenn Sie uns in dieser Studie unterstützen möchten, bitten wir Sie, den entsprechenden Abschnitt auf der nächsten Seite auszufüllen.

Falls Sie weitere Fragen zu dieser Studie haben, stehen wir jederzeit gerne für ein persönliches Gespräch zur Verfügung:

Prof. Dr. rer. nat. Michael Vorländer
Institut für Technische Akustik
RWTH Aachen
Kopernikusstraße 5, 52074 Aachen
Telefon: 0241-80-97985 | Email: post@akustik.rwth-aachen.de

Einwilligung zur Teilnahme am Hörversuch

„Wahrnehmung von Geräuschen elektrischer Maschinen"

Ich _____

(Name des Teilnehmers /der Teilnehmerin in Blockschrift)

bin schriftlich über die Studie und den Versuchsablauf aufgeklärt worden. Ich habe alle Informationen vollständig gelesen und verstanden. Sofern ich Fragen zu dieser vorgesehenen Studie hatte, wurden sie von Frau Samira Mohamaday vollständig und zu meiner Zufriedenheit beantwortet.

Mit der beschriebenen Erhebung und Verarbeitung meiner Leistungs- und personenbezogenen Daten bin ich einverstanden. Die Aufzeichnung und Auswertung dieser Daten erfolgt vollständig anonymisiert in der Professur für Technische Akustik der RWTH Aachen unter Verwendung einer Nummer und ohne Angabe meines Namens.

Sollten behandlungsbedürftige Auffälligkeiten im Hörscreening (Audiometrie) erkannt werden, bin ich damit einverstanden, dass mir diese mitgeteilt werden, so dass ich diese ggf. weiter abklären lassen kann. Ich wurde darüber informiert, dass die Information über auffällige Befunde u.U. mit versicherungsrechtlichen Konsequenzen verbunden sein kann.

Ich hatte genügend Zeit für eine Entscheidung und bin bereit, an der o.g. Studie teilzunehmen. Ich weiß, dass die Teilnahme an der Studie freiwillig ist und ich die Teilnahme jederzeit ohne Angaben von Gründen beenden kann. Ich weiß, dass ich in diesem Fall Anspruch auf eine Vergütung für die bis dahin erbrachten Stunden habe.

Eine Ausfertigung der Probandeninformation über die Untersuchung und eine Ausfertigung der Einwilligungserklärung habe ich erhalten. Die Probandeninformation ist Teil dieser Einwilligungserklärung.

Ich bin einverstanden, dass die von mir erhobenen Daten anonym in wissenschaftlichen Studien verwendet werden und deren Ergebnisse veröffentlicht werden.

_____

Name des/r Versuchsteilnehmers/in

_____

Ort, Datum, Unterschrift des/r Versuchsteilnehmers/in

Listening Test

Perception of Electric Engine Noise

Samira Mohamady, Michael Vorländer

Institute of Technical Acoustics (ITA), RWTH Aachen University

Description:

The listening test consists of two parts; Audiometry and Actual Test

Audiometry:

In this part the hearing abilities of each participant will be examined by using audiometry. The test will take place in an auditory room with audiometric Headphone to prevent any exterior interventions. We use a well-accepted method (bracketing) in which the participant listens to pure tone with different frequencies and sound pressures. The pulsation generates in some audible frequency range starting from low to higher. The level of each pulse increases with time, and the participant should push a button to indicate if he/she hears the sound, the button should be pushed as long as he/she hears the tone, and release it when the tone is not audible. This push-and-release method will determine the threshold of the participant at a certain frequency and this cycle will be repeated for some frequencies. This examination takes approximately 12-13 minutes.

Note: If any participant has a hearing loss above 20 dB, the person will be excluded from the actual test.

Actual Test:

The second part is the actual test. The test will take place in an auditory booth to prevent any exterior interventions. It is provided a monitor with keyboard and mouse inside the booth to perform the test. In the monitor screen, the listening test page is shown (Figure 1).

In the listening test screen (Figure 1) there are

➤ In the bottom-left: is start button to start the test (just push one time).

➤ In the upper side: three buttons that refer to the sounds to be played back; Reference Sound, sound A and sound B.

➤ In the middle part: there is an option to be activated if the subject wants to describe the sound differences with own words.

➤ In the bottom-right side: is the next button to continue the test for the next samples.

➤ In very low-left-bottom (red color font): is to abort the test (designed for administration purpose), the participants don't need to use this button.

➤ In the middle-bottom: question number is displayed, there will be in total 100 questions.

The participants are supposed to hear the Reference Sound, Sound A and Sound B and recognize sound which differs from the reference sound; the decision can be made by selecting the circles in the lower box of Sound A and Sound B. The participant could listen to the samples again (suggestion: try to hear the sounds not more than 2-3 times).

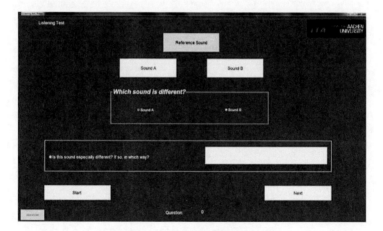

Figure 1. Main page of the listening test

➤ In the bottom-left: is start button to start the test (just push one time).

There will be a 30 second break in the middle of the test (after question no. 40); the system will automatically be disabled. At the same time the participant could evaluate the test, if it was hard or easy to differentiate.

The total test duration is estimated to be one hour. Participants will receive 8 € per each listening test.

More explanation will be provided in the day of the listening test.

A.4 Experimental Equipment

Table A.1: List of the main equipment

	Name	Producer	Quantity
1	Sound card and amplifier	ITA FireRobo	1
2	Microphones	KE4, Sennheiser	4
3	Omni-directional Loudspeaker	ITA	1
4	Accelerometer	B&K	1
5	Shaker	VISATON	1
6	Vibrational exciter	B&K, Type 4809	1
7	Hammer	ITA	1
8	Dummy Head	ITA	1
9	Head phone	Sennheiser HD650	1

Acknowledgments

I would like to express the deepest appreciation to Prof. Dr rer. nat. Michael Vorländer, Head of the Institute of Technical Acoustics (ITA) of the RWTH Aachen University, for welcoming me to his institute and for his support as my scientific advisor. I particularly thank him for his infinite patience, motivation, enthusiasm, and immense knowledge.

My thanks are also expressed to the secretary of the institute, Karin Charlier, for making my life easier. I also appreciate helpful talks with Dr Gottfried Behler, and express my gratitude for the academic support of Prof. Dr Ing. Janina Fels.

My sincere thanks go to the staff of the mechanical and electronic workshops at ITA, namely Rolf Kaldenbach, Norbert Konkol and Uwe Schlömer and all apprentices, for the outstanding work.

Many thanks to the helping hands that led to the conclusion of this thesis:

Dr Markus Müller-Trappet,

Dr Martin Guski,

Johannes Klein,

Dr Ellen Peng,

Lukas Aspöck,

Micheal Kohnen.

I would also like to thank all my colleagues and friends who made the institute a really pleasant working environment for me, in particular, my thanks go to Margret Engel and Fanyu Meng who were always great company, and whose help was invaluable during the challenging time of writing this study.

This thesis was made possible thanks to the generous financial support of the German Academic Exchange Service (DAAD), Germany.

Last but not least, I would like to acknowledge the unlimited love and support of my family, especially my mother and father. Many thanks also go to my lovely sister and her family and two supportive and wonderful brothers.

Bisher erschienene Bände der Reihe

Aachener Beiträge zur Technischen Akustik

ISSN 1866-3052

Alle erschienenen Bücher können unter der angegebenen ISBN-Nummer direkt online
(http://www.logos-verlag.de) oder per Fax (030 - 42 85 10 92) beim Logos Verlag
Berlin bestellt werden.